Spaceflight and the
Myth of Presidential Leadership

Spaceflight

and the Myth of Presidential Leadership

Edited by Roger D. Launius and Howard E. McCurdy

WITHDRAWN

University of Illinois Press

Urbana and Chicago

© 1997 by the Board of Trustees of the University of Illinois
Manufactured in the United States of America

I   2   3   4   5   C   P   5   4   3   2   I

This book is printed on acid-free paper.

Library of Congress Cataloging-in-Publication Data
Spaceflight and the myth of presidential leadership / edited by Roger D.
Launius and Howard E. McCurdy.
     p.   cm.
Includes bibliographical references and index.
ISBN 0-252-02336-6. — ISBN 0-252-06632-4 (pbk.)
     1. Astronautics—Government policy—United States. 2. Presidents—
United States. 3. Executive power—United States. I. Launius, Roger D.
II. McCurdy, Howard E.
TL789.8.U5S64   1997
387.8'0973—dc21
96-51213   CIP

Contents

Acknowledgments

To explore developments in U.S. space policy, the NASA History Office and the Center for Congressional and Presidential Studies organized a two-day symposium in the spring of 1993 that brought together government officials and experts on the American presidency. The symposium took place at the American University, where James A. Thurber, director of the Center for Congressional and Presidential Studies, was an early supporter of the project and deserves our thanks. Without his assistance this book could not have been completed. We wish to acknowledge the help of the staff of the NASA History Office: Patricia Shephard, who provided administrative support; Lee D. Saegesser, who helped track down documents; and J. D. Hunley, who read various drafts of the collected essays and provided valuable advice. In addition, we wish to acknowledge and thank Mark J. Albrecht, Giles Alston, Donald R. Baucom, Roger E. Bilstein, Rip Bulkeley, Tom D. Crouch, Philip Culbertson, Virginia P. Dawson, Duane A. Day, Henry C. Dethloff, Andrew J. Dunar, Tim Evanson, Aaron K. Gillette, Michael R. Gorn, Adam L. Gruen, R. Cargill Hall, Richard P. Hallion, James J. Harford, Ken Hechler, Gregg Herken, Jennifer M. Hopkins, Karl Hufbauer, Sylvia K. Kraemer, W. Henry Lambright, Pamela E. Mack, John E. Naugle, Allan A. Needell, Candice Nelson, Michael J. Neufeld, Arthur L. Norberg, John E. Pike, Willis H. Shapley, William S. Skerrett, Marcia Smith, Lawrence Suid, Joseph N. Tatarewicz, Stephen Waring, Glen P. Wilson, and Ray A. Williamson. All of these people would disagree with some of the areas chosen for emphasis and with many of the conclusions offered, but such is both the boon and the bane of historical inquiry. Since this work does not reflect the perspectives of all of these people, the editors retain responsibility for any errors of fact and judgment in the book.

Spaceflight and the
Myth of Presidential Leadership

Introduction:
The Imperial Presidency in the History of Space Exploration

Roger D. Launius and Howard E. McCurdy

The aggrandizement of presidential power during the administrations of John F. Kennedy, Lyndon B. Johnson, and Richard M. Nixon prompted a number of commentators to criticize the ease with which chief executives overwhelmed other centers of power in the United States. By the time of the Watergate affair, the expansion and abuse of presidential power relative to the Congress and courts had created a full-blown governmental crisis. Historians and political scientists such as Arthur M. Schlesinger Jr. decried the creation of what he called "the imperial presidency."[1] Like other commentators in the mid-1970s, Schlesinger feared that deference to the president had upset the traditional system of checks and balances.

This book deals with people who found relief rather than anxiety in the imperial presidency. Individuals who worked to promote the U.S. space program saw in the powerful presidency a solution to their most pressing problem—how to obtain the unfettered political support necessary to carry out projects like the expedition to the Moon in a political system that typically resisted long-range commitments. For space supporters, dynamic presidential leadership made big projects possible. For them, the imperial presidency was a godsend rather than a curse.

Concern over the imperial presidency among academics and political commentators was short-lived. Presidential power was in full decline by the administrations of Gerald R. Ford and Jimmy Carter in the latter half of the 1970s. Historians and political scientists, such as Thomas Cronin, issued tracts lamenting the gap between public expectations and presidential power.[2]

People promoting an aggressive space policy generally ignored these developments. Bewitched by Kennedy's 1961 commitment to send Americans to the Moon, they continued to profess their belief that visionary presidential leadership could overcome the obstacles created by constitutional checks and balances. Their faith in the ability of presidents to dominate the political system persisted long after outside commentators had concluded, in the words of Hugh Heclo, that presidential government was an illusion: "Presidential government is the idea that the president, backed by the people, is or can be in charge of governing the country. . . . This is an 'illusion' in the fullest sense of the word, for it is based on appearances that mislead or deceive."[3]

By examining the history of presidential leadership in the U.S. space program, this book reveals how the illusion of presidential government affected public policy. Not unexpectedly, the illusion created expectations that could not be satisfied. Space advocates pressed for the salvation that presidential leadership seemed to provide. Their faith in the capacity of presidential power to free them from the political thicket prevented them from adopting a more complex view of the forces affecting space policy. Not until the 1990s did this faith subside.

The ability of space exploration supporters to find salvation in the modern presidency drew its inspiration from Kennedy, who certainly had few illusions about the extent of presidential power when he delivered his speech challenging Americans to commit themselves "to achieving the goal, before this decade is out, of landing a man on the moon and returning him safely to earth." He announced that goal on May 25, 1961, as part of a speech before a joint session of Congress dealing with what he called urgent national needs brought on by the rigors of cold war rivalry with the Soviet Union. Kennedy departed extensively from his prepared text, pleading with congressional leaders to "consider this matter carefully . . . as it is a heavy burden." Kennedy added to his prepared remarks the following aside: "There is no sense in agreeing, or desiring, that the United States take an affirmative position in outer space unless we are prepared to do the work and bear the burdens to make it successful."[4] Kennedy knew that Congress could undercut his legislative initiatives by refusing to authorize them or, worse still, by authorizing his initiatives without appropriating the funds to carry them out. There is every reason to believe that he had a clear grasp of that political reality.

When Kennedy entered the White House, many people hoped that his inauguration would end the years of political deadlock most recently perpetuated by conflict between the Democratic Congress and the Republican president, Dwight D. Eisenhower. This did not occur. Kennedy's effort to break up the conservative coalition on the House Rules Committee succeeded by a mere five votes and only with the help of Republicans.[5] In the Senate, his own party refused to modify the filibuster rule, scuttling any hope Kennedy might have had for civil rights legislation in that session. Republicans provided the margin necessary to squeak Kennedy's emergency feed grains bill through the House, while conservatives rallied against his minimum wage bill by substituting a watered-down measure.[6] Kennedy's experience confirmed the words of the political scientist Clinton Rossiter, who had written during the midpoint of the Eisenhower administration that the president's tools for influencing Congress were "not one bit sharper than they were forty years ago."[7]

Neither party loyalty nor presidential prestige secured congressional support for Kennedy's measures. As Richard E. Neustadt had warned in 1960, the powers of the presidency amounted to little more than the power to persuade.[8] Kennedy was obliged to use his powers of persuasion to produce ad hoc coalitions for each new legislative initiative that he sent to Capitol Hill. Frustration over the lack of presidential power led the political scientist James McGregor Burns to publish *The Deadlock of Democracy* in 1967, in which he argued that congressional committee leaders constituted a separate political party independent of that which presidents employed to win elections.[9]

On the way back from Capitol Hill after his 1961 speech, Kennedy worried aloud about lack of enthusiasm for his space program proposals.[10] Based on the experience of the previous four months, he had good reason for concern. Proposals far more modest than the space initiative had encountered opposition from various sectors of the political spectrum. And Kennedy had not yet discovered at that time a reliable method for overcoming resistance.

The May 25 speech seemed to change that. Kennedy's space proposals sped through the Congress. The bill authorizing the buildup necessary to reach the Moon passed the Senate one month later on June 28. There was so little opposition that the senators did not even bother to take a recorded vote. The debate in the House was perfunctory, and the bill passed by a lopsided vote of 354 to 59. Kennedy noted the "overwhelming support by members of both parties" as he signed the bill authorizing his space initiatives on July 21.[11]

For many years, space boosters had searched for the key that would unlock the public treasury and provide them with the largesse necessary to undertake an aggressive program of space exploration. They had promoted space exploration through science fiction and popular astronautics. They had tied their dreams to the ballistic missile development movement, to the

International Geophysical Year, and to public fears about the cold war. They had received for this effort sufficient political support within the Eisenhower administration for only a modest program of satellite research and a single-seat Mercury capsule that only once spent more than a day in space.

With a single public declaration, Kennedy created a crash program to send humans to the Moon, as well as a supporting satellite and rocket program. In a remarkable departure from their prior behavior, other politicians deferred to Kennedy's goal. Congress did not undercut the initiative. The National Aeronautics and Space Administration (NASA) received the rarest of political commitments—eight years of uninterrupted support for a very public long-range science and technology endeavor. The speech in which Kennedy set the lunar goal remains one of the most memorable moments of that generation, in part because the results departed so dramatically from political norms. With such results, how could space enthusiasts not wish for an encore? Pundits applauded Kennedy's lunar commitment as well as his deft handling of other cold war crises, such as the 1961 Berlin showdown and the 1962 Cuban missile confrontation, more examples of presidential ability to act alone.

In fact, the political stage for Kennedy's dramatic lunar initiative was set well before he entered the presidency. His predecessor's parsimonious space program was motivated by a realistic desire to invest limited funds in military applications, such as reconnaissance satellites, rather than engage in what Eisenhower characterized as space stunts.[12] Democratic leaders, especially those in the Senate, recognized the symbolic importance of winning the space race in order to build trust in U.S. capabilities among people at home and abroad. After the launch of *Sputnik I* and *II* in late 1957, Majority Leader Lyndon B. Johnson urged Democratic senators to put aside budgetary restraints and engage the Soviets in the race into space. "Control of space means control of the world," Johnson warned.[13] Thus, politicians did not defer to Kennedy's decision so much as they endorsed what had already been decided.[14] Kennedy's moment is less dramatic in this context.

Afterward, the intellectual climate changed. In 1970, George Reedy, who had served as a special assistant to President Johnson, wrote a book in which he argued that the ability of presidents to rise above external dissent and criticism isolated them from the very forces designed to hold executives in check. "There is built into the presidency," Reedy argued, "a series of devices that tend to remove the occupant of the Oval Room from all of the forces which require most [people] to rub up against the hard facts of life on a daily basis."[15]

Reedy was responding to the actions of Presidents Johnson and Nixon, who seemed to behave more like monarchs than constitutional executives. In some instances, these presidents and their aides literally behaved as if they were above the law. The growth of presidential power, Schlesinger wrote in 1973, had produced "an unprecedented exclusion of the rest of the executive branch, of Congress, of the press and of public opinion" from decisions involving war and peace and the economy. Accordingly, the imperial presidency grew at the expense of other centers of power in the American polity. "Like the cowbird, it hatched its own eggs and pushed the others out of the nest," Schlesinger observed. "If this transformation were carried through, the President, instead of being accountable every day to Congress and public opinion, would be accountable every four years to the electorate. Between elections, the President would be accountable only through impeachment and would govern, as much as he could, by decree."[16]

All of this made for an odd situation within the network of space exploration enthusiasts. They had benefited from the use of presidential power and prerogative. People in the business of space exploration neither lamented the rise of presidential power nor deplored the ability of presidents to rule by decree. Instead, they asked for it to be done again. They implored Presidents Johnson and Nixon to endorse even more ambitious ventures. Space exploration advocates concluded that President Kennedy's ability to formulate a clear national commitment could be repeated, and in so doing would give the civilian space program the political protection it needed to turn general visions into engineering accomplishments.

Presidential leadership seemed even more imperative as the 1970s arrived. The scarcity of funds, the decline of superpower competition, and declining public interest in space all conspired to weaken the case for a strong space effort. These forces also weakened the ability of presidents to build consensus for a strong program, a point lost on too many space pioneers. Especially within NASA, strong presidential leadership was still viewed as the essential ingredient that would break the political logjam and allow leaders of the civilian space program to pursue their most ambitious goals.

The importance of presidential leadership had been impressed upon space promoters since the dawn of the space age in 1957. When Eisenhower proposed an exploration program that space enthusiasts viewed as excessively timid in the winter of 1957–58, the boosters appealed to both Congress and the White House. The House Space Committee attacked Eisenhower's proposals as a "beginner" program that lacked "proper imagination and drive."[17] In spite of congressional pressure for a more ambitious effort, led by personalities no less powerful than Senate Majority Leader Lyndon Johnson, Congress was unable to shake the administration from its plan.[18] What others, including members of Congress, could not do to bend Eisenhower, Kennedy seemingly did with one speech, creating the belief that the space program depended upon the willingness of the president to set expansive, idealistic, long-range goals.

This turn toward the executive guided virtually all subsequent efforts to establish goals in space beyond the landings on the Moon. Those efforts were attuned to an executive decision, with Congress apparently playing a secondary role. In 1969, President Nixon established a Space Task Group to advise "on the direction which the U.S. space program should take in the post-Apollo period."[19] Eleven years later, a transition team urged President Ronald Reagan to make a "definitive statement on space policy" at the earliest possible time. "A viable space program," the team members wrote, "must have purpose and direction." Without presidential leadership, they warned, the space program would "waste away."[20] Even Congress joined the clamor for executive leadership over the future space program. In 1984, Congress required the president to establish a National Commission on Space to map out future space policies.[21]

As the White House moved to occupy the center of space policy activity, the process for reviewing initiatives within the Executive Office of the President became more elaborate. Kennedy revived the National Aeronautics and Space Council, a White House body that Eisenhower had opposed, and made then Vice President Johnson its chair. The vice president acted as an intermediary to resolve disputes involving two or more agencies, especially NASA and the Department of Defense.[22] By the end of the 1960s, little important space policy making took place beyond the sphere of White House control, in spite of powerful congressional and bureaucratic interests that resided outside. Taking an approach that differed in form rather than substance, Nixon relied upon his Office of Management and Budget to analyze space issues and resolve interdepartmental disputes. The president, of course, retained final control over space policy decisions.[23]

As if in an effort to perpetuate the idea of executive leadership, the space policy apparatus within the White House grew more complex as presidential power declined. In 1982, President Reagan established the Senior Interagency Group for Space, a subcabinet council chaired by the assistant for National Security Affairs and empowered "to provide for orderly and rapid referral of space policy issues to the President for decisions."[24] In 1989, President George Bush recreated the National Aeronautics and Space Council to "oversee the implementation . . . of the president's space policy." By then, seven executive agencies as well as various members of the White House staff participated in the executive review process for space.[25]

As the machinery for executive decisions became more elaborate, the language of presidential commentaries about space became more definitive. In his May 1961 speech before Congress, Kennedy practically begged lawmakers to approve his initiatives on space. Eleven years later, Nixon's statement endorsing the space shuttle as America's next major initiative did not even mention the legislative branch. "I have decided today," Nixon announced from his presidential retreat in California, "that the United States should proceed at once with the development of an entirely new type of space transportation system."[26] Even though the star of the imperial presidency had fallen, Reagan was even less deferential when, like Kennedy, he appeared before a joint session of Congress to launch the nation's next major human spaceflight initiative. Reagan told lawmakers in 1984, "Tonight I am *directing* NASA to develop a permanently manned space station and to do it within a decade."[27] Only when Bush proposed in 1989 that the United States undertake a massive effort to return to the Moon and go on to Mars did the president acknowledge the growth in congressional power. Speaking from the steps of the Smithsonian's Air and Space Museum on the National Mall, he noted that America's future as a spacefaring nation would be decided on nearby Capitol Hill.[28]

While presidential power captivated space enthusiasts, scholars of the American executive observed its slow decline. "Few if any of our presidents have been the giants American mythology makes them out to be," Cronin explained in a book first published in 1975.[29] By overestimating the powers of the office, Cronin warned, people set up unrealistic expectations that would be soon disappointed. The ink was hardly dry on Schlesinger's *Imperial Presidency* when Congress repossessed the president's war-making powers, established a congressional budget process, and drove Nixon from office.[30] "As soon as the clamor over the 'imperial presidency' of Vietnam and Watergate subsided, the presidency appeared less conquering than conquered," the political scientist Aaron Wildavsky observed.[31] Space advocates failed to understand this development even as they sought to revive executive prerogatives by injecting international cooperation into space projects. In Washington, ideological and partisan divisions increasingly affected foreign affairs, and to many, Kennedy's presidency marked the dividing line. The experience that caused space buffs to worship at the altar of presidential power was apparently the top of the mountain. It had been all downhill since, a development lost on most people in the space policy network.

Illumination dawned slowly on the people who had tied their hopes to presidential prerogatives. Space exploration enthusiasts greeted Reagan's 1984 space station directive as a political mandate to take the "next logical step" in space. NASA established a schedule to produce a space station, as Reagan had directed, by 1994.[32] But nothing happened. Eight years after Kennedy offered his challenge, Americans stood on the Moon. Ten years after Reagan issued his directive, the space station was mired in political turmoil.

For exploration advocates, a full-fledged confrontation with reality occurred during the debate over the Space Exploration Initiative (SEI). In 1989, President Bush endorsed the ultimate spacefaring objective: human interplanetary travel. He proposed that the United States establish a lunar base and organize a human expedition to Mars, a decision on which he elaborated one year later: "Leadership in space takes more than just dollars: It also takes a decision. And so, I'm announcing one today. . . . I believe that before *Apollo* celebrates the 50th anniversary of its landing on the Moon the American flag should be planted on Mars."[33]

Outside the executive office, the presidential proposal was met with disbelief. In spite of a flurry of executive branch activity, Congress refused to appropriate even the modest funds necessary to study expedition technology. Bush complained that Congress "voted to pull the plug, completely gutting the seed money we proposed for the Moon/Mars mission." Commenting on the limits of presidential power, he observed: "Space used to be a bipartisan effort: an American effort. . . . Unfortunately, not everyone on Capitol Hill shares this commitment to investing in America's future."[34]

The demise of SEI, concurrent with continuing troubles for the earth-orbiting space station, at last forced space exploration champions to question long-held assumptions about presidential omnipotence. Faith in the ability of presidential commitments to free space policy from the constraints of Washington politics finally declined. As the journalist Leonard David pointed out, expensive space efforts such as SEI are extremely tough sells.[35]

Twenty years earlier, President Nixon had told space supporters to expect as much. Thomas O. Paine, NASA administrator in 1968–70, had put relentless pressure on Nixon to make a commitment to the agency's post-Apollo goals, telling him a month after his inauguration that he had to make a "general directive to define the future goals of manned space flight in the next few months."[36] Nixon was more realistic, and he suggested in 1970 that Americans stop thinking about space activities "as a series of separate leaps, each requiring a massive concentration of energy." He added: "Space expenditures must take their proper place within a rigorous system of national priorities. What we do in space from here on in must become a normal and regular part of our national life."[37] Space ceased to be special by the time the first Americans reached the Moon.

Most space supporters did not understand how truly exceptional the Apollo mandate was. After the glamor of Kennedy's moment dimmed, space policy came to rest alongside all the other priorities of government for which presidential leadership played a diminishing role. This eventually disappointed those who believed in the power of presidents to make space exploration special. The Apollo decision was, therefore, an anomaly in the history of the U.S. space program.[38]

In reality, questions of space policy and the specific projects that have developed from it are but a microcosm of larger political trends that have affected the entire government. The rise and fall of presidential power in American politics as a whole can be observed in detail within the development of the U.S. space program. Seven essays on presidential space policy are gathered here. They resulted from the two-day symposium organized in the spring of 1993 by the NASA History Office and the Center for Congressional and Presidential Studies. These essays analyze presidential leadership and its relationship to the many agencies of the executive branch, Congress and its staff, special groups outside of the government, and the American public. Each essay views the U.S. space effort as a policy issue as well as a scientific, technical, and engineering endeavor.

The symposium sought to bring together thoughtful scholars and senior government officials in an atmosphere conducive to an honest review of the U.S. space effort. Leading scholars of the American presidency participated in the symposium. None of them—and this was an important criteria for their participation—had written in any detail about the space program before, and they were therefore able to comment on it from the larger perspective of public policy and presidential leadership. This fresh perspective contributed to an overall reassessment of the role of the president in defining and directing the space program.

Five of the essays in this volume deal with specific presidencies and the approach of each to the development of U.S. space policy. David Callahan and Fred I. Greenstein continue the revision of Dwight D. Eisenhower's reputation as president, which has been underway for more than a decade. They argue that the image of Eisenhower as an amiable "do-nothing" president, who smiled and played golf while crises threatened, is incorrect. Eisenhower worked strenuously behind the scenes while giving the appearance of inaction, and in most instances his indirect approach to leadership was highly effective. He used the power of the emergent "imperial presidency" to establish a modest effort that took a measured approach toward space, while doing so in an inconspicuous way.

Michael R. Beschloss's essay on John F. Kennedy and the decision to go to the Moon reinforces the view that the early 1960s were the high point of presidential power in formulating space policy. Using a wealth of documentary information, Beschloss notes that Kennedy's 1961 announcement came at a crucial time in U.S. history when the president could exert control over cold war activities with a relatively free hand. Even as public interest in the cold war waned, the Apollo decision remained a model for a generation of space promoters who saw it as the best means for continuing a far-reaching and assertive exploration agenda.

Robert Dallek's essay on Lyndon B. Johnson and the politics of space comments on how Johnson used both his presidential office and his unparalleled knowledge of Congress to ensure that Apollo was completed within the decade as the fallen Kennedy had proposed. At the same time, Johnson refused to endorse any new, large-scale space endeavors. Both actions reinforced belief in the power of the president to create an invincible space program. By the time Johnson left office, space exploration advocates were firmly committed to the idea of the "imperial presidency" as the only sure means for preserving their future space exploration plans. Advocates did not understand, Dallek makes clear, the difficulties Johnson experienced in maintaining a coalition of interests in support of Project Apollo and how he had to use divergent arguments to sell it to a reluctant Congress.

Joan Hoff's provocative essay on the space program under Richard Nixon and his successors in the 1970s attacks head-on the faith of space program advocates in the power of the presidency. Nixon refused to endorse a strenuous follow-on effort to Project Apollo but did so without convincing space program leaders that his support would mean little in the new social and political environment emerging from the late 1960s. Nixon's inaction further reinforced the belief among space enthusiasts that the president was strong enough to make new exploration goals a reality provided he could be convinced of their legitimacy. Rather than accommodate themselves to the new policy formulation process, space supporters placed the blame on the personality of the president and his unwillingness to step up to the kind of greatness that they believed Kennedy had exhibited.

Lyn Ragsdale's chapter describes how two Republican presidents of the 1980s, Ronald Reagan and George Bush, invoked the rhetoric of Kennedy and wholeheartedly endorsed an aggressive space program. At the same time, their policies never received strong political support from Congress, from competing sectors of the federal bureaucracy, or from the public at large. Only during the Bush administration did space exploration advocates begin to see that the idea of an "imperial presidency" was a myth. As Ragsdale shows, the twin political failures of the early space station program and the Space Exploration Initiative prompted space policy enthuasists to alter their perspectives on the role of presidential leadership in favor of one more attuned to the realities of modern government.

Robert H. Ferrell examines the impact of international relations on the formulation of U.S. space policy. Among space advocates, the notion that international affairs strengthens the president's hand is a central article of faith. Cold war competition certainly strengthened Kennedy's ability to win support for the lunar objective. Subsequent programs, however, have served a variety of purposes within U.S. foreign policy. Ferrell shows how international considerations have driven U.S. space policy, both enlarging and constraining the latitude for choice.

John M. Logsdon explores the relationship between the desire for presidential leadership and the use of the space program to assist in achieving a position of national supremacy. Finally, the editors of the volume conclude with a basic commentary on NASA's search for an alternative paradigm to shape its space policy agenda. Since Kennedyesque leadership statements have been a chimera in the agency's history, what forces do shape space policy political coalitions? The editors examine the effect of political partisanship, basic ideology, and "pork-barrel" politics on the national space agenda.

Taken together, this collection of essays provides an analysis of the relationships among the president, Congress, and the bureaucracy in formulating and conducting space policy. The essays emphasize the idea of the "imperial presidency" and the misguided reliance on presidential edicts to forward space exploration agendas. The essays demonstrate the way in which political ideas can outlast the realities that create them. The Apollo decision and its accomplishment under Kennedy and Johnson blinded leaders of the civilian space effort to emerging realities and made it hard for them to adjust to a different environment. Since that brief moment in the 1960s when a president prevailed, space exploration advocates have been forced to wrestle with policy questions in a far different manner.

Notes

1. Arthur M. Schlesinger Jr., *The Imperial Presidency* (Boston: Houghton Mifflin, 1973).

2. Rexford G. Tugwell and Thomas E. Cronin, eds., *The Presidency Reappraised* (New York: Praeger, 1974); Thomas E. Cronin, *The State of the Presidency* (Boston: Little, Brown, 1975); Hugh Heclo and Lester M. Salamon, eds., *The Illusion of Presidential Government* (Boulder, Colo.: Westview Press, 1981); Aaron Wildavsky, *The Beleaguered Presidency* (New Brunswick, N.J.: Transaction Publishers, 1991).

3. Heclo and Salamon, *Illusion of Presidential Government*, p. 1.

4. Kennedy, "Presidential Address on Urgent National Needs," May 25, 1961, *Public Papers of the Presidents of the United States: John F. Kennedy, 1961* (Washington, D.C.: Government Printing Office, 1962), pp. 401–5.

5. This is discussed in James N. Giglio, *The Presidency of John F. Kennedy* (Lawrence: University Press of Kansas, 1990).

6. "Agriculture," *Congressional Quarterly Almanac, 1961* (Washington, D.C.: Congressional Quarterly, 1961), p. 79.

7. Clinton Rossiter, *The American Presidency*, rev. ed. (New York: New American Library, 1962), p. 240.

8. Richard E. Neustadt, *Presidential Power* (New York: John Wiley, 1960).

9. James McGregor Burns, *The Deadlock of Democracy* (Englewood Cliffs, N.J.: Prentice-Hall, 1967).

10. John M. Logsdon, *The Decision to Go to the Moon: Project Apollo and the National Interest* (Cambridge, Mass.: MIT Press, 1970), p. 129.

11. "Budget Summary," *Congressional Quarterly Almanac, 1961*, p. 424.

12. Eisenhower, "Why I Am a Republican," *Saturday Evening Post*, Apr. 11, 1964, p. 19.

13. Johnson, "Statement to the Democratic Conference," Jan. 7, 1958, LBJ collection, box 23, Lyndon Baines Johnson Library, Austin, Texas.

14. See Robert A. Divine, *The Sputnik Challenge: Eisenhower's Response to the Soviet Satellite* (New York: Oxford University Press, 1993).

15. George Reedy, *The Twilight of the Presidency* (New York: New American Library, 1970), p. 17.

16. Schlesinger, *Imperial Presidency*, pp. 208, 377.

17. National Advisory Committee for Aeronautics to Killian's Office, White House, Aug. 6, 1958, NASA Historical Reference Collection.

18. Logsdon, *Decision to Go to the Moon*, pp. 35, 95–100.

19. Space Task Group, *The Post-Apollo Space Program: Directions for the Future* (Washington, D.C.: Executive Office of the President, 1969), pp. 27–29.

20. George M. Low, Team Leader, to Richard Fairbanks, director, Transition Resources and Development Group, "Report of the NASA Transition Team: National Aeronautics and Space Administration," Dec. 19, 1980, pp. 7, 39, NASA Historical Reference Collection.

21. Public Law 98–361, 98th Cong., 2d sess. (July 16, 1984), title II.

22. Logsdon, *Decision to Go to the Moon*, p. 70.

23. Logsdon, "The Decision to Develop the Space Shuttle," *Space Policy* 2 (May 1986): 103–19.

24. Reagan, "United States National Space Policy," *Weekly Compilation of Presidential Documents,* July 4, 1982, p. 875; Howard E. McCurdy, *The Space Station Decision: Incremental Politics and Technological Choice* (Baltimore: Johns Hopkins University Press, 1990), pp. 139–42.

25. Bush, "Message to Congress Transmitting a Report on the Establishment of the National Space Council," *Weekly Compilation of Presidential Documents,* Feb. 1, 1989, p. 271.

26. Nixon, "Space Shuttle Program," *Weekly Compilation of Presidential Documents,* Jan. 5, 1972, p. 27.

27. Reagan, "State of the Union," *Weekly Compilation of Presidential Documents,* Jan. 25, 1984, p. 90, emphasis added.

28. Bush, "Remarks by the President at 20th Anniversary of Apollo Moon Landing," July 20, 1989, NASA Historical Reference Collection.

29. Cronin, *State of the Presidency,* 2d ed. (Boston: Little, Brown, 1980), p. 2.

30. Richard P. Nathan, *The Plot that Failed: Nixon and the Administrative Presidency* (New York: Wiley, 1975), pp. 70–76; David McKay, *Domestic Policy and Ideology: Presidents and the American State, 1964–1987* (Cambridge: Cambridge University Press, 1989), p. 65.

31. Wildavsky, "The Party of Government, the Party of Opposition, and the Party of Balance: An American View of the Consequences of the 1980 Election," in *The American Elections of 1980,* ed. Austin Ranney (Washington, D.C.: American Enterprise Institute, 1981), p. 330.

32. NASA, Office of Space Station, *The Space Station: A Description of the Configuration Established at the Systems Requirements Review (SSR)* (Washington, D.C.: Technical and Administrative Services Corporation, June 1986).

33. Bush, "Remarks at the Texas A&M University Commencement Ceremony in Kingsville, Texas," *Weekly Compilation of Presidential Documents,* May 11, 1990, pp. 749–50.

34. White House, Office of the Press Secretary, "Text of Remarks by the President at the Marshall Space Flight Center," June 20, 1990, NASA Historical Reference Collection.

35. Leonard David, "Space Report: Planning Missions to the Moon and Mars," *Ad Astra,* Dec. 1990, pp. 16–20. See also Logsdon, "Looking for Leadership," *Ad Astra,* July/Aug. 1990, pp. 10–14.

36. Paine to Nixon, "Problems and Opportunities in Manned Space Flight," Feb. 26, 1969, NASA Historical Reference Collection. The Bureau of the Budget recommended to the president that he resist Paine's entreaty and opt for reductions in NASA's budget. See Robert P. Mayo to Nixon, "Proposed Budget Amendment for the Space Program," Mar. 3, 1969, RG 51, series 69.1, Office of Management and Budget, box 51–78–31, National Archives and Records Administration, Washington, D.C.

37. Nixon, "The Future of the United States Space Program," *Weekly Compilation of Presidential Documents,* Mar. 7, 1970, p. 329.

38. This has been demonstrated too many times to be seriously questioned. See Walter A. McDougall, . . . *The Heavens and the Earth: A Political History of the Space Age* (New York: Basic Books, 1985), pp. 141–235; Logsdon, *Decision to Go to the Moon;* Harvey Brooks, "Motivations for the Space Program: Past and Future," in *The First 25 Years in Space: A Symposium,* ed. Allan A. Needell (Washington, D.C.: Smithsonian Institution Press, 1983), pp. 3–26; Rip Bulkeley, *The Sputniks Crisis and Early United States Space Policy: A Critique of the Historiography of Space* (Bloomington: Indiana University Press, 1991).

1

The Reluctant Racer:
Eisenhower and U.S. Space Policy

David Callahan and Fred I. Greenstein

QUESTION: Mr. President, the burden of some recent statements
on Capitol Hill, primarily by generals, has been that we are well
behind the Russians in missile development, with little or no prospect of catching up with them in the near future. I'd like to ask
you, sir, as far as man's effort to enter space, as well as the development of military missiles, do you feel any sense of urgency in
catching up with the Russians?

THE PRESIDENT: I am always a little bit amazed about this business of catching up. What you want is enough, a thing that is
adequate. A deterrent has no added power, once it has become
completely adequate, for compelling the respect of any potential
opponent for your deterrent and, therefore, to make him act
prudently.[1]

The story of Dwight D. Eisenhower and U.S. space policy is that of a reluctant participant in a highly public program of research and development
that had all of the earmarks of a race but that the participant himself res-
olutely defined as a nonrace. It is in part a story of technological competition, but in a larger sense it is a story of political competition—that is, both
partisan, national competition between a popular president and a congres-
sionally based coalition of members of the opposite party and a cold war
of international competition between the United States and the Soviet
Union. It is also a story of the reluctance of a president to invoke the presidential office to mandate an aggressive space program. In that sense, Eisenhower used the power of the emergent "imperial presidency" to hold back
what he considered reckless actions in the face of a cold war crisis.

During the 1950s, Eisenhower was widely seen as a presidential figure-
head who depended on his staff for policy direction and day-to-day decision making. Today, it is scarcely news to scholars that Eisenhower was

in fact very much the architect and principal constructor of the policies and actions of his administration.[2] In Eisenhower, the United States had a president who was far more politically shrewd and able than was evident to most of his contemporaries. And he was as much a geopolitical strategist as a politician. As a two-term cold war president, Eisenhower brought a remarkably unified and, in the judgment of recent analysts, coherent strategic stance to his conduct of national security.[3]

Space policy during the 1950s provides an ideal case study of the strengths and weaknesses of Eisenhower's leadership style as it has come to be known in the years since reexamination of his presidency became an intellectual growth stock. It is an excellent example of how initial negative assessments of Eisenhower's actions have been modified or abandoned with the passage of time and the declassification of information. It provides, also, a fascinating contrast with the direction space policy was to take under Eisenhower's successors.

The hallmark of Eisenhower's handling of space policy was his stolid resistance to demands that the United States embark on crash programs to compete with the Soviet Union. To understand this measured approach, it is instructive to consider certain of the individual qualities of the man, as well as the broad strategic stance of his administration and the state of U.S. space policy prior to *Sputnik*. This sets the stage for a detailed examination of the policies and actions of the Eisenhower administration following the Soviet space launching of October 4, 1957. That event was Pearl Harbor–like in the extent to which it galvanized the American people and their leaders, leading to a fundamental redirection of the nation's policies and priorities.

This chapter will focus on Eisenhower himself and the distinct imprint his vision of national security issues placed on space policy in the 1950s. In so focusing, it is necessarily selective, building on the work of other scholars who have analyzed the complex interplay of political and military considerations and the intense bureaucratic and partisan maneuvering that characterized space policy in the Eisenhower years. As Eisenhower's statement in the epigraph of this chapter suggests, the politics of space in the 1950s was in many ways subordinate to the politics of military missile development. The concern here, however, is not mainly with missile policy and the missile-gap controversy but with space policy and the space-gap controversy.[4]

Although the story of Eisenhower and space policy unfolds for the most part in the 1950s, Eisenhower lived through the first months of the Nixon presidency, remaining alert and preoccupied with contemporary affairs almost to his dying day. His views from the sidelines, which we consider in our conclusion, are of interest not only for their own sake but also for the insight they shed on counterfactual questions about how space policy might have unfolded had Eisenhower's policies been continued into the 1960s.

Space Policy before *Sputnik*

A starting point for any discussion of space policy in the 1950s must be a recognition of how intimately linked this issue was with broader national security concerns. Both before and after *Sputnik*, the prevalent view among U.S. government officials was that space represented a challenging new forum for cold war competition. Eisenhower, more than any other public figure of the time, resisted this notion. To understand from whence this resistance grew, it is necessary to understand Eisenhower's views on national security.

Eisenhower's National Security Philosophy

Dwight D. Eisenhower entered the White House with a more fully articulated view of national security policy than any president before or since. His interest in the broad questions of security and strategy went back to his tutelage under the legendary military intellectual General Fox Conner in the early 1920s. Ike had served as supreme commander in Europe during World War II and Army chief of staff and supreme allied commander (SAUCER) of NATO forces in the postwar period. Eisenhower had more than just a professional's factual knowledge in the defense area; his firm convictions about domestic as well as foreign policy, and the relationship between them, comprised a full-fledged philosophy of national security. "Spiritual force, multiplied by economic force, is roughly equal to security," Eisenhower wrote to Lucius Clay in 1952. "If one of these factors falls to zero, or near zero, the resulting product does likewise."[5]

On domestic policy, Eisenhower was a free-market conservative. He believed that big government and high taxes were the great enemies of prosperity. As he constantly reminded those around him, one of his chief missions at the White House was to contain the growth of government expenditures. Eisenhower fervently believed that budgets should be balanced and frequently warned about the perilous consequences of not achieving this goal.[6] He adamantly resisted the view of economists such as the former chairman of President Truman's Council of Economic Advisors, Leon Keyserling, who held that higher government spending could stimulate the economy and thus generate new revenues that made up for any deficits. During a 1955 press conference, Eisenhower commented that he had read that "Mr. Keyserling has a plan for spending a good many more billion dollars, for reducing taxes, and balancing the budget at the same time. That, I would doubt, was a good economic plan."[7]

This conservatism, along with a strategic doctrine that rejected the need for overkill, had a direct impact on Eisenhower's thinking about the defense budget. "How to balance essential security needs with maximum economic strength was the great equation that Eisenhower strove to solve," Ivan Morgan has written.[8] In his first message to Congress, Eisenhower warned that boosting military strength "without regard to our economic capacity would be to defend ourselves against one kind of disaster by inviting another." On April 30, 1953, Eisenhower was told by the National Security Council (NSC) that the United States faced two fundamental threats: the external Soviet menace and the internal danger that the costs of defending the free world "may seriously weaken the economy of the United States and thus destroy the very freedom, values and institutions which we are seeking to maintain."[9]

This message was a centerpiece of Eisenhower's national security thinking, preached to both the public and his own advisors. "Again and again I reiterated my philosophy on the defense budget: Excessive spending causes deficits, which cause inflation," Eisenhower wrote in his memoirs. "Every addition to defense spending does not automatically increase military security. Because security is based upon moral and economic, as well as purely military strength, a point can be reached at which additional funds for arms, far from bolstering security, weaken it."[10]

Beyond his fear of the economic consequences of excessive federal spending, Eisenhower had a Republican distrust of government. He worried that a larger government could undermine democracy by producing a bureaucratic monolith that was accountable to no one. As time passed, Eisenhower became particularly concerned about the growing influence of military and scientific elites. He voiced this concern most strongly, of course, in his farewell address when he warned against the "acquisition of unwarranted influence, whether sought or unsought, by the military-industrial complex."[11] There is evidence that Eisenhower harbored these concerns from early in his White House tenure.

In assembling his cabinet, Eisenhower turned to people who shared his concern about the overall damage to America's position that could be wrought by high government spending. Eisenhower's closest economic advisor, Secretary of the Treasury George M. Humphrey, was a strong believer in restrained government spending, lower taxes, and balanced budgets. "Humphrey's fiscal views reflected his conviction that many government activities were wasteful, unnecessary and the harbinger of socialistic collectivism," Morgan observed. Humphrey was an especially harsh critic of defense spending, saying at one point in 1957, "We're throwing away forty billion in capital every year—on the dump heap." He added that it "serves only our security for that year, then on the dump heap."[12] Eisenhower's secretary of state, John Foster Dulles, was apprehensive about overly zealous attempts to save money at the Pentagon but generally adhered to the administration line. "If economic security goes down the drain, everything goes down the drain," Dulles affirmed.[13]

While Eisenhower saw economic peril in every budget increase and worried about democracy's future in a technocratic world, he was less concerned than many of his contemporaries about the Soviet threat. As supreme commander of NATO forces, Eisenhower had pondered the Soviet threat on a daily basis. The experience seems to have left him less, not more, concerned about the prospect of bold Soviet aggression. While in the White House, Eisenhower never put credence in the idea that the Soviets would mount an attack at the first sign of Western weakness. On one occasion in 1953, he complained to his special assistant for national security affairs, Robert Cutler, that members of the National Security Council "worry so damn much about what we'll do when the Russians attack. . . . Well, I don't believe for a second they will ever attack."[14] On another occasion, in 1956, Eisenhower commented in a letter to Field Marshal Bernard Law Montgomery about Soviet intentions: "These Communists are not early Christian martyrs. The men in the Kremlin are avid for power and are ruthlessly ambitious. I cannot see them starting a war merely for the opportunity that such a conflict might offer their successors to spread their doctrine."[15]

During the late 1950s, a time that some strategic thinkers such as Albert Wohlstetter and Paul H. Nitze advertised as a period of "maximum danger," Eisenhower remained confident about the U.S. security position. James P. Killian Jr., Eisenhower's first science advisor, remembered the president getting up from the chair in his office, looking out the window, and talking about his own experience as a general. Eisenhower said that he hoped his advisors recognized that he had some measure of judgment in this field and that he didn't see any possibility of hostilities with the Soviet Union. Killian also recalled Eisenhower telling him that he himself was not "anticipating or expecting any shooting war with the Soviet Union for the next five years."[16]

Beyond his conviction that the Soviets would not risk initiating war in the nuclear age, Eisenhower firmly believed that the West as a whole was distinctly stronger than the communist world and would remain so given its superior economic performance. In 1951, when top Truman administration officials were warning of the West's disintegrating position vis-à-vis the Soviet Union, Eisenhower stated: "We must not forget that in total wealth, material strength, technical scientific achievement, productive capacity, and in rapid access to most of the raw materials of the world, we, the free nations, are vastly superior to the communist bloc."[17] Eisenhower repeated this belief often during his presidency and reiterated it with particular frequency following the Soviet launch of *Sputnik*. His clear message was that quantitative analyses of military hardware conveyed only part of the story—and a very small part at that—about America's security situation.

The "New Look" and Early Space Policy

The Eisenhower administration's economizing approach to national security was exemplified by its "New Look" defense policy. The New Look offered a return to the Truman administration's miserly defense spending approach that had reigned prior to the Korean War, and it rejected the wartime articulation of the 1950 cold war planning document, NSC 68, which responded to that Asian crisis. In a speech before the Council on Foreign Relations in January 1954, Dulles enunciated the Eisenhower administration's objections to Truman's post-1950 cold war strategy, saying it could not have been sustained for long "without grave budgetary, economic, and social consequences."[18]

In concrete terms, the New Look translated into a greater emphasis on nuclear weapons for defense, including ballistic missile development, and reduced spending for conventional forces. The overall effect of the policy was to reduce the expansion but not the overall growth of defense spending. In fiscal year (FY) 1954, defense expenditures constituted 65.7 percent of the federal budget and 12.8 percent of the gross national product (GNP). By FY 1961, such expenditures had dropped to 48.5 percent of the budget and 9.1 percent of the GNP.[19]

In the crisis following *Sputnik,* critics of the Eisenhower administration erroneously charged that his New Look program of austerity had served to undermine both ballistic missile research and the development of a U.S. satellite. Through unimaginative leadership and penurious policies, critics charged, Eisenhower had left the United States at a distinct disadvantage in the opening round of the space race. Ironically, it was the pre-1950

Truman policies relative to defense spending that had retarded U.S. missile development.[20]

The notion that space was a sphere for international competition predated the Eisenhower presidency. As early as 1946, some experts had warned about the negative consequences of falling behind in the space race. A RAND report written in that year suggested that the nation that first put a satellite into space would be seen as militarily and scientifically superior. It predicted massive consternation if the United States found that another nation had beat it in putting up a satellite. A report commissioned by the Truman administration in 1952 echoed this finding, arguing that a Soviet advantage in satellites would be a serious blow to U.S. scientific prestige and would be milked by Soviet propagandists for all it was worth.[21]

Eisenhower was ambivalent about the issue of prestige in the cold war. Prestige was a relatively minor factor in his broad-based conception of Western strength and the nature of cold war competition, but he was intensely interested in propaganda and psychological warfare. Believing that psychological warfare was a cost-effective way to score cold war gains, Eisenhower emphasized it from the earliest days of his administration, devoting both personal attention and budgetary resources to bolstering America's propaganda activities abroad. Indeed, psychological warfare was discussed at Eisenhower's first cabinet meeting on January 23, 1953. Within his first year in office, Eisenhower had reorganized the U.S. propaganda apparatus, creating a new Operations Coordinating Board (OCB), which had psychological warfare as one of its missions and would eventually involve itself heavily in U.S. space activities.[22]

In 1954, U.S. space policy began to take shape with planning for the International Geophysical Year (IGY), July 1, 1957, to December 31, 1958. During that year, Wernher von Braun of the Army Ballistic Missile Agency wrote a report in which he argued that putting a satellite into space was eminently feasible. Braun argued that, since this goal could be realized by the United States in only a few years with available technology, "it is only logical to assume that other countries could do the same. It would be a blow to U.S. prestige if we did not do it first."[23]

Von Braun's view was echoed the following year in NSC 5520, a government directive on space policy that was approved by Congress on May 20, 1955. The document recognized the feasibility of orbiting a civilian satellite and stated, "Considerable prestige and psychological benefits will accrue to the nation which first is successful in launching a satellite. The inference of such a demonstration of advanced technology and its unmistakable relationship to intercontinental ballistic missile technology might have important repercussions on the political determination of free world countries to resist communist threats, especially if the USSR were to be the first to establish a satellite."[24]

Nelson A. Rockefeller, then the special assistant to the president on government operations and vice chairman of the OCB, circulated NSC 5520 through the government with a cover memo of his own. The successful launching of a satellite, he wrote, will "symbolize scientific and technological advancement to peoples everywhere. The stake of prestige that is involved makes this a race that we cannot afford to lose."[25]

With congressional approval of NSC 5520, the U.S. civilian satellite program, Project Vanguard, was officially born. However, this enterprise was not conducted with the urgency that Rockefeller's warning might have warranted. Prestige had been only one of four main reasons listed in NSC 5520 for developing a civilian satellite; it was not touted as the chief motivating factor. Just as important were military research considerations and the desire to establish a legal precedent during the IGY for satellite overflight of foreign countries, along with a drive toward scientific achievement.

In short, during this initial, pre-*Sputnik* stage of the U.S. space program, there was no consensus in the federal government for waging an outright competition with the Soviet Union to reap the psychological dividends of being first into space with a civilian satellite. Eisenhower himself seems to have been unconcerned with winning such a competition in 1955–56. At an NSC meeting on May 3, 1956, where the escalating cost of Vanguard was discussed, Eisenhower acknowledged that he had never been very enthusiastic about the satellite program. He rejected suggestions by Treasury Secretary Humphrey that the program be cancelled on economic grounds but said that the priority assigned to Vanguard should be below that of more urgent Pentagon programs. Eisenhower's stance, as summarized in the minutes of the May 3 meeting, was that the United States should continue its program to launch a satellite with the understanding that the program "will not be allowed to interfere with the ICBM and IRBM programs but will be given sufficient priority by the Department of Defense in relation to other weapons systems to achieve the objectives of NSC 5520."[26] In January 1957, Eisenhower was told that the first attempt at a satellite launch was scheduled for October 31, 1957. He did not object to this timetable. As Eisenhower later wrote in his memoirs: "Since no obvious requirement for a crash satellite program was apparent, there was no reason for interfering with the scientists and their projected time schedule."[27]

If Eisenhower was relatively unconcerned about losing a prestige race in space, he was by no means complacent when it came to the military applications of missile technology and the intelligence potentiality of satellites. In the summer of 1954, he asked MIT President James Killian Jr. to head a commission to examine current trends in the military competition with the Soviet Union and to evaluate the threat of surprise attack. The recommendations of Killian's Technological Capabilities Panel (TCP), put forth in February 1955, would have an important impact on U.S. space policy over the next several years. First, and most importantly, the TCP recommended that the Air Force program for ICBM development be given the highest priority. Eisenhower approved this recommendation and, as Killian would later write, this was the "first time such a priority had been given in peacetime."[28] With a special "missile czar," Assistant Secretary of Defense Donald Quarles, who coordinated the effort in the Pentagon, the U.S. missile program was essentially run on a crash basis through the rest of the decade. As Eisenhower later recalled, "To these programs we devoted all the resources that they could usefully absorb at any given time."[29]

The effect of this priority status for military missiles, however, was to delay the U.S. civilian satellite project. As Killian observed, Vanguard's development was "handicapped by the National Security Council Directive that gave the development of our military missiles top priority with the result that many able engineers working on Vanguard were diverted to ICBM programs."[30] In the wake of *Sputnik*, the Eisenhower administration defended itself by observing that the United States could have put a satellite in orbit before the Soviets but that such an effort would have hurt top priority missile programs. Vanguard has "not had equal priority with that accorded our ballistic missile work," said a White House statement released shortly after the Soviet launch. "Speed of progress in the satellite project cannot be taken as an index of our progress in ballistic missile work."[31]

Another result of the Technological Capabilities Panel was to draw attention to the need for better U.S. intelligence capabilities. "We must find ways to increase the number of hard facts upon which our intelligence estimates are based, to provide better strategic warning, to minimize surprise in the kind of attack, and to reduce the danger of gross overestimation or gross underestimation of the threat," said the report.[32] This recommendation echoed a 1954 RAND report that argued that developing a satellite reconnaissance vehicle was of vital importance. On March 16, 1955, the Air Force took an initial step toward this goal when it called for proposals from industry to create a U.S. spy satellite. This project, too, would take precedence over the civilian science satellite.

The first U.S. reconnaissance satellite was not operational until 1960. In the meantime, starting in June 1956, the United States relied on the U-2 spy plane program to gather intelligence on Soviet military capabilities.[33] Over the next several years, U-2 flights revealed that the Soviet missile program was proceeding very slowly. In part, it was Eisenhower's access to this information that explained his confident outlook during the furor following the Soviet *Sputnik* launch.

In his classic history of the space age, Walter A. McDougall succinctly summarized the complicated history of U.S. space policy during the first half of the 1950s:

> Occupied by the need to keep abreast of the USSR in long-range rocketry, the Eisenhower administration put the ICBM on a crash basis. Absorbed by the need to monitor Soviet R & D and deployment whether arms race or arms control obtained, it also gave priority to the USAF spy satellite program, two and one-half years before the Space Age opened. Worried about the legal and political delicacy of satellite overflight, it seized the IGY opportunity to initiate an unobtrusive scientific satellite program under civilian auspices. Finally, the administration was advised of the propagandistic value of being first into space. Of all these critical areas, however, the last had the lowest priority.[34]

With more generous funding there is no reason why the United States could not have pursued all three of its main space programs on a top priority basis. However, to accept a case for such funding Eisenhower would not only have had to suspend his perpetual resistance to higher defense spending but also to have become convinced that the warnings about the danger to U.S. prestige by a Soviet first in space had sufficient merit to warrant a more costly American space program.

Such warnings never resonated strongly with Eisenhower. Still, it would be wrong to conclude that he never worried about losing the race to put a civilian satellite into space. Slightly under five months before the launching of Sputnik, at a May 10, 1957, meeting of the NSC, Eisenhower expressed concern that efforts to make the Vanguard satellite more scientifically sophisticated would delay the program. Such costly instrumentation had not been envisioned when NSC 5520 had originally been approved, Eisenhower noted. He stressed that "the element of national prestige, so strongly emphasized in NSC 5520, depended on getting a satellite into orbit, and not on the instrumentation of the scientific satellite."[35]

These concerns were expressed too late to change the course of the program. In June, after statements by Soviet scientists that the Soviet Union would soon launch a satellite, the Operations Coordinating Board began preparing the Eisenhower administration's response to losing the first round of the space race. Central to that response, agreed members of an OCB working group, should be a disclaimer by the United States that it ever had any intention of engaging in a race with the Soviets to launch the first civilian satellite.[36]

Sputnik: Its Impact and Immediate Aftermath

The Soviet launch of Sputnik I touched off one of the most serious crises of Eisenhower's presidency. Like no other previous event, it cast doubt on his capacity for decisive presidential leadership and undermined his strongest asset: a reputation for sound judgment in the national security field.[37] Eisenhower responded to the Soviet challenge with confidence and steadiness, but these personal characteristics were at once an asset and a handicap. On the one hand, a more insecure president might have overreacted to the crisis, authorizing unproductive crash programs to counter the Soviet move or making belligerent pronouncements about America's determination and ability to win the space race. Responses like these could have heightened cold war tensions. On the other hand, Eisenhower appears not to have appreciated just how panicked Americans were or to have recognized the degree to which space could become politicized. To some extent, this seeming complacency reflected Eisenhower's mistrust of rhetoric and his insufficient appreciation for the symbolic importance of policy. To a greater extent, it reflected his confidence in America's security position.

In the aftermath of the Soviet launch, Eisenhower sought to contain a number of consequences that he found distressing: the perception among both the public and certain elites of a new sense of military vulnerability, which contrasted sharply with Eisenhower's own outlook; the widespread tendency to see space as a new arena of cold war competition, which Eisenhower believed was misguided; and the rapid manner in which space policy became politicized by Democrats who found the alleged space and missile gaps perfect issues for attacking the Eisenhower administration without personally attacking the popular president.

The Immediate Impact of Sputnik

News of the Soviet launch of Sputnik on October 4, 1957, stunned Washington and the nation. In the tense climate of cold war competition, even minor jolts to the politico-military equilibrium could be nerve-racking. But Sputnik was a decidedly major jolt. It appeared to signal both a broad Soviet technological superiority and, more ominously, a specific Soviet advantage in ballistic missiles. Sputnik was the greatest propaganda coup of the cold war, and it triggered a torrent of alarmed comment. Senator Henry Jackson called *Sputnik* "a devastating blow to the prestige of the United States as the leader of the free world."[38] Senator Lyndon Johnson and others compared the Soviet satellite launch to Pearl Harbor.[39] Newspaper editorials around the country warned of America's eroding position vis-à-vis the Soviet Union.

In the next few months, with the Soviet launch of a second, far more impressive Sputnik satellite in early November and the highly publicized explosion on the launchpad of America's Vanguard satellite in December, Eisenhower faced unrelenting criticism on the space issue. Even though the United States succeeded in launching its own satellite in January 1958 and rapidly organized an impressive space program, the perception of a lagging U.S. space effort dogged Eisenhower for the rest of his time in office. The space-gap issue, moreover, remained hopelessly intertwined with fears of U.S. military vulnerability, fueled in 1958–60 by increasingly strident allegations that the United States was yielding the advantage in the cold war.

While Eisenhower and his top advisors were caught unprepared for the extraordinary national and international uproar that followed *Sputnik,* they were not altogether surprised that the Soviets had managed to launch a satellite. The U-2 spy plane had taken photos of the SS-6 missile on which *Sputnik* would be launched, and U.S. intelligence had told Eisenhower in November 1956 that the Soviets would be able to launch a satellite within a year. As William Burrows observed, "By the time Sputnik went into orbit on October 4, the United States knew quite a bit about the missile that carried it there."[40] Apparently, U.S. intelligence was not entirely comprehensive, for in his memoirs Eisenhower wrote that he and others were astounded by the weight of the Soviet satellite, 184 pounds. "The size of the thrust required to propel a satellite of this weight came as a distinct surprise to us."[41]

What startled Eisenhower far more than the advance in Soviet rocketry was the intensity of public concern.[42] *Sputnik* was not true proof of a Soviet advantage in ICBM development, but it *appeared* to be—and this idea was terrifying to many in the United States. Killian, who would be appointed White House science advisor in November 1957, captured the furor of the moment in his memoir:

> As it beeped in the sky, Sputnik I created a crisis of confidence that swept the
> country like a windblown forest fire. Overnight there developed a widespread
> fear that the country lay at the mercy of the Russian military machine and that
> our government and its military arm had abruptly lost the power to defend the

homeland itself, much less to maintain U.S. prestige and leadership in the international arena. Confidence in American science, technology, and education suddenly evaporated.[43]

If Eisenhower was indeed out of touch with this national panic, part of the reason undoubtedly was his own lack of alarm. As Killian later wrote: "With his full knowledge of our military programs, especially our progress in missile and military satellite technology, and our national intelligence estimates, he found it hard to understand the national dismay and fear. He was startled that the American people were so psychologically vulnerable."[44] Because he believed America remained secure, Eisenhower did not think that *Sputnik* should be allowed to trigger sweeping changes in national policy. He acknowledged the need, as he recalled later, for the United States to "take all feasible measures to accelerate missile and satellite programs."[45] Yet for the most part he felt his chief problem was a political one—that of convincing the American people that all was well and that their nation remained not only secure but actually superior to the Soviet Union in overall strength.

Eisenhower's way of tackling this challenge was to seek to educate the public about the facts of national security as he saw them. Although Eisenhower has been criticized by historians for his failure to appreciate the power of the bully pulpit, his public relations effort following *Sputnik* was quite vigorous. It was sustained over time and hewed to a consistent message. In an October 9 press conference, Eisenhower insisted that the *Sputnik* launch did not raise his apprehension "one iota. I see nothing at this moment, at this stage of development, that is significant in that development as far as security is concerned."[46]

Eisenhower observed that the Soviet Union had still not substantiated its claim that it possessed an accurate, operational ICBM. Other administration officials echoed this reassuring theme. In a speech in San Francisco on October 15, Vice President Richard Nixon insisted that "militarily, the Soviet Union is not one bit stronger today than it was before Sputnik was launched." He added that the free world "remains stronger than the Communist world" and could "meet and defeat any potential enemy."[47] Dulles made the same point in remarks to the press a day later.

But the administration was fighting an uphill battle in the face of the all-too-visible evidence of Soviet achievements, including the more impressive *Sputnik II*, which was launched on November 3, 1957, with a 1,121-pound payload, including a dog. *Sputnik II* not only underscored the power of Soviet missile boosters but also provided evidence that the Soviets were already striving toward manned spaceflight.

Eisenhower's most substantial effort to quell the near hysteria that followed the Soviet launches came in a major television and radio address on November 7. Again, Eisenhower assured the public that America's nuclear arsenal was adequate to deter any Soviet threat. He could not, of course, reveal to the public the intelligence he was receiving from U-2 flights over the Soviet Union—data that showed the Soviet missile program to be still in a state of infancy. But there were many other reassuring points he could and did convey. He explained the elaborate warning system the United States had to protect against any surprise attack and talked about how dispersal of the U.S. strategic arsenal made it invulnerable to a Soviet first strike. He acknowledged that the Soviets were likely ahead in some missile areas and in satellite technology, but he assured his listeners that, overall, "we are well ahead of the Soviets in the nuclear field both in quantity and in quality. We intend to stay ahead."[48]

Paralleling Eisenhower's message that America was winning the arms race was his emphatic insistence that the United States was not engaged in a space race. Eisenhower had already made this point in his October 9 statement, saying, "The United States satellite program has been designed from its inception for maximum results in scientific research. . . . Our satellite program has never been conducted as a race with other nations."[49] During his November 7 speech, he reiterated this point. Over the next three years, he continued to emphasize the noncompetitive nature of the U.S. space program.

Eisenhower's effort to avoid a highly publicized space race was motivated not only by cost considerations and fear that the United States might lose such a race because of its late start but more fundamentally by geopolitical considerations. Since the beginning of his administration, Eisenhower had sought to contain the competition with the Soviet Union. He believed that the cold war struggle represented a colossal waste of human resources. He also believed, as noted earlier, that the more intense that struggle became, the more America's democratic institutions and way of life would be threatened.

To Race or Not to Race: Eisenhower Deliberates

Even as he publicly dismissed the notion of a space race, Eisenhower privately expressed concerns about the prestige and propaganda dimensions of space policy. At a meeting with scientific and military advisors on October 8, 1957, he agreed with a suggestion that the Defense Department consider using the Jupiter-C missile as a backup to Vanguard to ensure the success of U.S. efforts to get a satellite into space as soon as possible. The

Jupiter-C missile had been proposed by the Army's Redstone Arsenal as the vehicle to launch an IGY satellite in 1955, but it had been rejected in favor of the Navy's Project Vanguard, largely because the Navy's candidate did not interfere as much with higher priority ballistic missile programs and it seemed to offer greater scientific return. Later in October, the Pentagon officially began planning for a Jupiter launch in early 1958.[50]

At a National Security Council meeting on October 10, Eisenhower was briefed on Project Vanguard and told that the U.S. satellite would orbit at a lower altitude than *Sputnik*. Eisenhower's response was to question whether such a lower orbit might affect U.S. prestige. Later at the same meeting, according to the minutes, "the President stressed once again the great political and psychological advantage of the first achievement of an IRBM and an ICBM. He noted that from the inception of the ballistic missile program the Council had agreed that these political and psychological considerations were perhaps even more important than the strictly military considerations."[51]

Eisenhower's thinking throughout his presidency and retirement was marked by the tension between his conviction that space exploration should not be the subject of international competition and his realization that it inevitably was. He mused in a January 1958 meeting with his party's congressional leaders on the irony that "we should undertake something in good faith only to get behind the eight-ball in a contest which we never considered a contest."[52] In 1965, he explained to a letter writer that "under no circumstances did we want to make the thing a competition, because a race always implies urgency and spectacular progress regardless of cost. . . . Neither then nor since have I ever agreed that it was wise to base any of these projects on an openly and announced competition with any other country. This kind of thing is unnecessary, wasteful and violates the basic tenets of common sense." Yet, in the same letter Eisenhower commented that "manifestly we did not want to be second in the field."[53]

Eisenhower's concern about prevailing in the ostensible noncompetitive space race increased over time, but he continued to confine expressions of such concern to private meetings. By 1959, Eisenhower was dwelling frequently on the need for the United States to speed up its development of a large booster missile, or superbooster, which he saw as having tremendous psychological significance.[54] And while publicly Eisenhower continued to emphasize that increased scientific knowledge was the main goal of the U.S. space program, privately he began to rank that goal last—behind the goals of national security and prestige. At a meeting with top advisors on October 21, 1959, for example, Eisenhower emphasized three goals for the space program: "The first is that we must get what Defense really needs in space; this is mandatory. The second is that we should make a real advance in space so that the United States does not have to be ashamed no matter what other countries do; this is where the super-booster is needed. The third is that we should have an orderly, progressive scientific program, well balanced with other scientific endeavors."[55]

By 1960, the aim of avoiding shame loomed large in Eisenhower's mind. Thus, at a January 12, 1960, NSC meeting, he declared that the United States should seek to "achieve a psychological advantage for ourselves," adding that "we would have to eliminate" whatever discrepancy existed between America and the USSR and "in certain instances would have to exceed Soviet accomplishments."[56]

Eisenhower continued to believe personally and stress publicly that American prestige was rooted most firmly in U.S. economic success and that a crash space program to bolster America's image was neither necessary nor desirable. But clearly, between late 1957 and 1960 his views underwent an evolution. Two factors appear to have changed Eisenhower's thinking: first, the clear concern with prestige on the part of the Soviet leadership and, second, the emergence of a strong consensus within the United States that success or failure in space policy was integral to the nation's world standing.

Nikita Khrushchev's frequent emphasis on the psychological component of the cold war was impossible to ignore. Whatever Eisenhower's doubts, the Soviet leader manifestly believed that prestige mattered in the superpower rivalry, and Khrushchev sought to gain maximum political leverage from Soviet gains in space. Even before the *Sputnik* launches, Khrushchev had exaggerated Soviet progress in developing ballistic missiles and touted Soviet science generally. His aim was to intimidate U.S. allies, to woo newly decolonized developing countries by advertising the superiority of the communist economic system, and (it is now known) to obscure major Soviet military weakness.

After the *Sputnik* launchings, Khrushchev stepped up his propaganda effort, boasting about the devastation that could be wrought in Western Europe by Soviet nuclear strikes and citing the Soviet satellites as proof of the Soviet Union's scientific prowess. Eisenhower may not have been easily shaken by such posturing, but from the first days of the *Sputnik* crisis many of his advisors showed intense concern about the propaganda implications of space exploration. At the NSC meeting on October 10, 1957, CIA Director Allen Dulles commented that Khrushchev "had moved all of his propaganda guns in place. The launching of an earth satellite was

one of a trilogy of propaganda moves, the other two being the announcement of a successful testing of an ICBM and the recent test of a large-scale
hydrogen bomb at Novaya Zemlya." Dulles claimed that the Soviet propaganda offensive was aimed at creating maximum leverage in the Middle East and, more generally, at demonstrating the effectiveness of the
communist system to the underdeveloped countries. In Dulles's view, the
campaign was "exerting a very wide and deep impact."[57]

Other U.S. officials shared this view. At the same meeting, Undersecretary of State Christian Herter described the overseas reactions to *Sputnik*
as "pretty somber" and argued that the United States "will have to do a
great deal to counteract them and, particularly, to confirm the existence
of our own real military and scientific strength." Arthur Larson, head of
the U.S. Information Agency, echoed this point, saying, "If we lose repeatedly to the Russians as we have lost with the earth satellite, the accumulated damage would be tremendous." Larson insisted that the United States
must be first in achieving the next big breakthrough in space.[58]

In the immediate aftermath of the *Sputnik* launches, U.S. officials were
unsure about the international ramifications of the Soviet achievements in
space. The State Department and Central Intelligence Agency (CIA) received a flood of reports of reactions from around the world, and sorting
through this information took time. On November 14, Gordon Arneson,
the deputy director of Intelligence and Research at the State Department,
summarized the preliminary view of some U.S. analysts regarding these
reactions in a memorandum to Secretary Dulles:

> The USSR's prestige has risen substantially and the U.S. has suffered a serious,
> although not decisive, setback. . . . World opinion tends to hold that the sput-
> niks per se have not altered the strategic balance of forces in the short run, since
> Soviet ICBMs are not yet thought to be in mass production. Nevertheless, some
> new weight has been lent to Soviet foreign policy pronouncements and increased
> credibility may attach to Soviet claims in other fields.

Arneson saw few immediate consequences of this new credibility but went
on to express a view that was quickly becoming conventional wisdom
within the Eisenhower administration: "Delayed or insufficient demonstration of United States success in the ballistic field would produce political
and psychological effects of substantially more serious nature—for example, on attitudes toward neutralism and on the cohesion of alliances."[59]

Outside of the Eisenhower administration, there was a widespread belief that the *Sputnik* launches had pushed the cold war rivalry into a new
arena. In congressional hearings in late 1957, a parade of expert witnesses echoed the judgment of Dr. Vannevar Bush, wartime head of the Office

of Scientific Research and Development and the most prestigious scientist in the United States at the time, who commented, "In the scientific field we must recognize that we are in a tough competitive race with the Russians and have a lot of good tough work to do."[60] Even at this early stage, some members of the scientific/technological elite speculated that a principal arena for U.S.-Soviet competition would be a race to the Moon.

In February 1958, the RAND Corporation produced a report that analyzed the political implications of the space age. The report argued that the developments in space could have far-reaching implications. It would "be folly to deny that the allies' estimates of the balance of power in the future are based in part on the expectation that Western science and technology will maintain a decisive lead over the Soviet bloc." Such perceptions were closely linked to space exploration, and competition in this field had to be managed with an eye to propaganda gains. "From now on, the U.S. should recognize the need for restoring credibility in U.S. superiority, stress our peaceful intentions and their aggressive ones, and disclose and publicize U.S. outer space activities according, first and foremost, to the effect on the U.S. international position."[61]

Also in February, Eisenhower's science advisory committee produced a paper on space policy that stated, "The psychological impact of the Russian satellites suggests that the U.S. cannot afford to have a dangerous rival outdo it in a field which has so firmly caught, and is likely to hold, the imagination of the world."[62] This conclusion was reflected in a document approved by Eisenhower in August 1958, NSC 5814/1, "Preliminary U.S. Policy on Outer Space."[63] Within the next two years, other space policy documents called for an unequivocal U.S. victory in space. Implicit in this policy-making process was a fear that failure to catch up with the Soviets might give rise to the idea that the United States was now "second best." A chief U.S. objective, therefore, should be "to achieve and demonstrate an overall superiority in outer space without necessarily requiring U.S. superiority in every phase of space activity."[64]

Still, the public face of U.S. space policy remained noncompetitive. A widely disseminated 1958 White House statement on space policy, "Introduction to Outer Space," stated that strength in space technology "will enhance the prestige of the United States among the peoples of the world and create added confidence in our scientific, technological, industrial, and military strength." But the statement as a whole paid almost no attention to the competitive aspects of space exploration, dwelling instead on the scientific wonders of venturing beyond the earth's atmosphere.[65]

In the final analysis, while Eisenhower did come to worry more about the connection between prestige and space policy as time passed, he and his closest advisors on space remained ambivalent on this point. T. Keith Glennan recalled, for example, that, during the private meeting in which Eisenhower offered him the top NASA job in August 1958, Ike "made no mention of any great concern over the accomplishments of the Soviet Union although it was clear that he was concerned about the nature and quality of scientific and technological progress in this country."[66] In a 1959 memorandum to Eisenhower, Glennan wrote: "Personally, I do not believe we can avoid competition in this field. . . . But I do believe that we can and should establish the terms on which we are competing. We could thus place the 'space race' in proper perspective with all the other activities in the competition between the US and USSR."[67]

Killian, probably Eisenhower's most influential advisor on space policy, also believed the United States should walk this fine line. As he said shortly after leaving the White House in 1959 to return to the Massachusetts Institute of Technology: "I believe that in space exploration, as in all other fields that we choose to go into, we must never be content to be second best, but I do not believe that this requires us to engage in a prestige race with the Soviets. We should choose our own objectives in space science and exploration and not let the Soviets choose them for us by copying what they do. . . . In the long run we can weaken our science and technology and lower our international prestige by frantically indulging in unnecessary competition and prestige-motivated projects."[68] As longtime Eisenhower aide General Andrew Goodpaster recalled, Eisenhower shared such views: "The President's approach was if we're doing the right thing in about the right way we'll let the prestige work itself out."[69]

The Domestic Politics of Space

Besides trying to head off an outright race, Eisenhower sought to quell the partisan bickering that surrounded space policy after *Sputnik*. This, too, proved difficult, and Eisenhower was subject to more criticism on missile and space policy than in nearly any other area during the course of his presidency. Leading the attack were Democrats in Congress who hoped to improve their party's prospects in the 1960 presidential and congressional elections. The Democrats suggested that American inferiority in missile and space achievements vis-à-vis the Soviet Union underscored a broader failure by Eisenhower and other Republican leaders to provide sound national leadership.[70] Indeed, the putative space and missile gaps became part and parcel of the larger Democratic stance in the 1960 presidential campaign, and the urgency of revising such deficiencies was central to much of Kennedy's more inspirational rhetoric.

It is testimony to Eisenhower's discomfort about politicizing national security policy that he refused to try to exculpate himself by blaming the Truman administration for its slow pace in missile development. While Eisenhower hinted in some of his speeches that Truman was to blame for America's late start in space, and made this point explicitly during the 1958 congressional campaign, he did not fully express his true feelings on this point until he published his memoirs in the early 1960s.[71] In *The White House Years: Waging Peace,* Eisenhower quoted his 1947 remark as Army chief of staff that a neglect of research on guided missiles "could bring our country to ruin and defeat in an appallingly few hours." He then noted that in the seven years of FY 1947–53 the United States programmed less than $7 million for long-range ballistic missiles. On two separate occasions, the executive branch failed to spend money that Congress had appropriated to the Air Force for this purpose. Eisenhower recalled that once in the White House he immediately set out to reverse this pattern of neglect.

Another point that Eisenhower made in his memoirs but did not stress while president was his view that the Democratic-controlled Congress shared much of the blame for deficiencies in U.S. space policy. He argued that Congress had slowed down Vanguard in the first half of 1957 by interfering with Pentagon efforts to use emergency funds for the project. In addition, he expressed annoyance at members of Congress who had threatened to reduce the Defense Department budget by $2 billion.[72] All of these arguments could have been made by Eisenhower as president in response to the criticism that was heaped on him after *Sputnik,* but only at the cost of further politicizing space policy and undercutting his own opposition to a crash space program.[73]

The *Sputnik* crisis, along with the recession that had begun in August 1957, ushered in a period in which Eisenhower was no longer invulnerable to criticism. Between January and November 1957, his popularity plummeted from 79 percent approval to 57 percent approval, according to a Gallup poll.[74] Eisenhower's mild stroke in late November did not help matters. "The long honeymoon was over," wrote Robert Divine. "For five years Eisenhower had presided over a period of peace and prosperity, basking in public gratitude for ending the Korean war and letting the nation enjoy a great material abundance. Now he suddenly had to convince a skeptical nation that he understood the new problems facing the country but that he possessed the energy and vision needed to restore the United States to its accustomed position of world primacy."[75]

The Unfolding of Space Policy, 1958–61

Following the *Sputnik* launches, there was little question that the United States would pursue a stepped-up program for space exploration. Space was a frontier that could not be ignored, and perhaps more widespread than the feeling of fear among the American public in the wake of *Sputnik* was an intense curiosity about space. The sale of books and magazines that dealt with space and rockets soared, as did membership in clubs and associations in these areas. Eisenhower himself was intrigued if not convinced of the necessity of a large-scale, expensive program. Even as a general, well before the missile age, Eisenhower had expressed his belief in the likelihood of future space travel.[76]

Though Eisenhower was no space buff, his science advisor, Killian, for one, felt that the president nonetheless had a distinct personal interest in space exploration. Killian saw this interest as rooted in Eisenhower's broader appreciation for the importance of science. Killian went so far as to compare Eisenhower to Thomas Jefferson, suggesting that there was "an interesting parallel between Jefferson's scientific interests and Eisenhower's intellectual hospitality to those he called 'my scientists,' and to scientific and technological matters."[77]

Beyond the basic conviction that America would have a larger space program after *Sputnik,* there was substantial uncertainty in late 1957 and early 1958 about exactly what the goals of this program would be, how it would be organized, and the amount of money it would cost. Eisenhower resolved this uncertainty by seeing to it that the early space program was relatively modest, that it would be clearly separated from the military drive to develop ballistic missiles and reconnaissance satellites, and that the organizational setup for space exploration would be an independent civilian agency able to resist vested interests and military domination. Finally, Eisenhower's inevitable aim was to restrain spending on space, in keeping with his overall desire to check the growth of the federal budget.

Cartoonists of the time depicted Eisenhower as napping or golfing while the Soviets gained the advantage in space. But in truth he was closely involved in mapping out a carefully circumscribed American agenda for space activities that sought to exploit its military potential and to further scientific knowledge.

Space and National Security Policy

After *Sputnik*'s launch, the issues of space exploration and national security were inextricably linked in the minds of many. Eisenhower faced pressures to increase defense spending from within the government as well as from the Democrats. The most intense pressure of this kind came from a body of national security experts that Eisenhower himself had convened, the Security Resources Panel—or Gaither Committee, so named for its chair, H. Rowan Gaither, prominent attorney and member of the board of the Rand Corporation. The Gaither Committee had been established in mid-1957 to analyze U.S. civil defense needs. It had soon broadened its mandate to include the entire gamut of strategic issues. *Sputnik* was launched as the committee was completing its work, and that event helped to solidify the view of top members that the United States was fast falling behind in the arms race.[78] In particular, the Soviet launch seemed to add weight to predictions made by such defense analysts as Albert Wohlstetter that the United States would soon be vulnerable to a preemptive Soviet nuclear strike. In its final form, presented to Eisenhower on November 7, 1957, the Gaither Report advocated a drastic increase in U.S. military preparations.[79]

Eisenhower rejected most of the report's recommendations for new military spending; he believed that the committee's assessment of U.S. strategic vulnerability was greatly exaggerated. But politically, the timing of the report—the essence of which soon leaked to the press—could hardly have been worse. At precisely the moment that Eisenhower was seeking to reassure the American public that *Sputnik* had little significance, a group of respected experts had raised the specter of a widening missile gap. To many observers, the connection between the Soviet Union's new preeminence in space and America's endangered security appeared self-evident. And nowhere was the zeal for stressing this link greater than on Capitol Hill, where Democratic members of Congress repeatedly invoked Soviet gains in space in calling for a major step-up of U.S. defense efforts. The successful launch of U.S. satellites by early 1958 did nothing to quiet administration critics.

Despite such pressures, Eisenhower held the line, rejecting the allegation that Soviet successes in space meant impending superiority in arms. Between FY 1958 and FY 1960, defense expenditures actually declined as a percentage of both the GNP and the federal budget.[80]

The Origins of NASA

While Eisenhower was determined to keep the issues of space and security politically separated, he was also committed to separating the areas organizationally. Initially, Eisenhower did not see the need for a separate agency for space exploration. At a February 4, 1958, meeting, Killian told Eisenhower that many in Congress were pressing for some space work to be done outside of the Defense Department. Eisenhower responded that he did not think that large operating activities should be put in another organization because of the duplication. He also worried that putting talent into crash programs outside of defense would undermine the higher priority missile programs. Eisenhower indicated that his condition for allowing the Defense Department to continue handling space initiatives was that it "gets its own organization correct, i.e., that there is a central organization to handle this in Defense."[81]

Eisenhower's initial inclination to keep the space program as part of the Defense Department was consistent with his general desire to restrain the growth of government. Taking space out of the military's hands would mean creating a new bureaucracy, a prospect Eisenhower could not have relished. He may also have hoped to avoid a fight with the Pentagon, which opposed the creation of a separate agency for space exploration and had big plans for space-related undertakings. Whatever his initial reasoning, Eisenhower soon changed his mind and came to favor civilian control of space exploration. Explaining this shift in his memoirs, he wrote, "Information acquired by purely scientific exploration could and should, I thought, be made available to all the world. But military research would naturally demand secrecy."[82] In effect, Eisenhower came to see that two space programs would be better than one: a vigorous military space program would receive top priority and spearhead America's missile and spy satellite programs; a civilian program would be the public face of American space exploration, undertaking those operations that had only propagandistic or scientific value. Such a division of labor exists to this day.

The process by which Eisenhower handled the organizational aspects of space policy, establishing NASA, reflected his strong faith in his science advisors and his desire to rise above politics. By late 1957, intense competition was under way in Washington among various bureaucratic players for control of space exploration. The two main contenders were the Defense Department and the National Advisory Committee for Aeronautics (NACA), a federal research agency formed in 1915 "to supervise and direct the scientific study of the problems of flight, with a view to their practical solution." NACA became an enormously important government research and development organization for the next half century, materially enhancing the development of aeronautics in the United States. The fight over space looked like it might be as messy as the battles over atomic energy in the late 1940s.[83]

Eisenhower approached this fray by stepping away from it and depoliticizing his decision to the greatest degree possible. He turned the problem of organizing a space program over to Killian and the President's Science Advisory Committee (PSAC), asking in early February that it recommend the outlines of a space program and the organization to manage it.[84] Eisenhower's decision on this point is characteristic of his hidden-hand approach to leadership. He knew well that Killian shared many of his views on science, space, and the cold war competition.

Killian had backed a civilian space agency since late 1957. And well before Eisenhower formally asked for a recommendation on space organization, PSAC's position was that an enlarged NACA should oversee civilian space missions. As one scholar observed in an early investigation of NASA's birth: "PSAC, vocally representing the interests of the scientific community, sought a primarily civilian structure in which basic research and peaceful space missions could be pursued free from military control."[85] Later, explaining his own enthusiasm for NACA, Killian wrote, "Here was a government scientific agency that was under the lay direction of some of the best civilian talent in the country, and the organization operated with freedom from political influence and unencumbered by government bureaucracy and red tape."[86] NACA itself was more than willing to take on the mission of space exploration, lobbying actively for the assignment. Thus, wrote Enid Curtis Bok Schoettle, "by the end of January, the group of scientific advisers whom Eisenhower had charged with designing a space program and the agency's leadership were agreed that NACA would be the base on which NASA would be built."[87]

The idea quickly won widespread support inside the executive branch. On March 5, 1958, Eisenhower approved a memorandum ordering the Bureau of the Budget to draft a bill for Congress that would turn NACA into the National Aeronautics and Space Administration (NASA). The draft was completed by late March, and on March 27, Eisenhower said, "I expect to send up shortly recommended legislation providing for civilian control and direction of governmental activities incident to a civilian space program."[88] After intensive debate and tinkering, the bill establishing NASA was approved by Congress and signed into law on July 29, 1958.

In reflecting later on Eisenhower's relationship with his science advisors, Killian observed that the president "turned to our group repeatedly for advice when he felt that recommendations reaching him on military or other matters were colored by special interests."[89] The creation of NASA was the foremost example of Eisenhower's reliance on PSAC to sort out fiercely conflicting claims. "This whole undertaking is a vivid example of what can be accomplished by a group of advisers, freed by the president of bureaucratic controls and wearing the president's mantle," Killian stated.[90]

At the February 4 meeting in which Eisenhower discussed the organization of the space program, he had said that he did not want to concern himself with the details of the problem.[91] By turning the matter over to PSAC, Eisenhower succeeded in this goal, and there is no evidence that he anguished personally over how to organize space policy. While Killian felt he was caught in a "political hurricane,"[92] Eisenhower seems to have felt only a strong breeze. This episode is a classic example of Eisenhower's leadership style: he had gotten exactly the outcome he wanted without appearing to engage in any outright political maneuvering.

Eisenhower's choice of Glennan as first NASA administrator served further to point U.S. space policy in the president's preferred direction. As with Killian, Glennan—then head of Case Institute of Technology in Cleveland—was a highly respected and independent figure who happened to share Eisenhower's basic outlook toward science and space.

Glennan described his attitude toward his new job in a memoir he wrote after leaving NASA. First, he believed, like Eisenhower, that government was "growing too large" and that every effort should be made to avoid "excessive additions to the Federal payroll." Second, he was concerned that the United States proceed at the right pace—"orderly but aggressive"—in an area that was filled with technological uncertainty. Third, he shared Eisenhower's view that the prestige value of space exploration could not be ignored, nor should the competition with the Soviets dictate America's space program. "In effect," wrote Glennan, "this meant that we must avoid the undertaking of particular shots, the purpose of which would be propagandistic rather than directed toward solid accomplishments in understanding the environment with which we are dealing."[93]

The Space Race Accelerates

Given the political pressures for an all-out space race with the Soviet Union, the degree to which Eisenhower controlled the space policy agenda in the late 1950s stands as a considerable achievement. If Eisenhower had genuinely been the passive president many of his contemporaries supposed, he would never have achieved such control and instead found himself buffeted by public opinion, outmaneuvered by powerful congressional leaders, and manipulated by his own bureaucracy. Eisenhower suffered none of these fates in the area of space policy.

It would be inaccurate, however, to suggest that he was ever really in command of events. Eisenhower was correct in his claim that under his watch the United States "deliberately avoided hysterically devised crash programs and propaganda stunts" in space. But despite claims to the contrary, both at the time and in later years, early U.S. space policy was indeed heavily determined by what the Soviet Union did, especially in the years 1959–60. The most significant indication of this was the initiation of Project Mercury, the program to put a man in orbit around the earth.[94]

In the wake of *Sputnik II,* it had become clear that the next major milestone in space exploration would be to place a human being in space. PSAC's 1958 report, *Introduction to Outer Space,* had identified manned flight in orbit as an obvious and attainable goal of space exploration. The same report had speculated about the requirements for a manned lunar landing. The administration's first major policy statement on space, NSC 5814/1, approved in August 1958, had also cited the inevitability of manned space exploration and elucidated its political consequences. The report argued that the "time will undoubtedly come when man's judgment and resourcefulness will be required fully to exploit the potentialities of outer space." Manned flight, it suggested, could have a major impact on world politics even greater than *Sputnik.* "No unmanned experiment can substitute for manned exploration in its psychological effect on the peoples of the world." NSC 5814/1 predicted the Soviets would be able to mount such a flight by 1959–60.[95]

The implication of this prediction was clear: if the United States wanted to have any chance of avoiding what Killian said could be "a recurrence of the Sputnik hysteria if the Soviets get a 'man in space' first,"[96] it had to initiate a major program to beat them into orbit. By September 1958, a special panel on manned flight declared the U.S. goal was to "achieve at the most early practicable date orbital flight and successful recovery of a manned satellite."[97]

Project Mercury represented everything Eisenhower claimed he wanted to avoid in space policy. It was hugely expensive, driven almost entirely by competition with the Soviets, and lacked a compelling scientific rationale. A 1960 PSAC report on putting a man in space resorted to inspirational language, declaring that "among the major reasons for attempting the manned exploration of space are emotional compulsions and national aspirations. These are not subjects which can be discussed on technical grounds." The panel concluded that "man-in-space cannot be justified on purely scientific grounds, although more thought may show that there are situations for which this is not true."[98] Glennan later commented about

Mercury: "As one looks back on that decision, it is clear that we didn't know very much about what we were doing."[99]

Eisenhower's approval of Project Mercury paralleled his failure to control NASA's budget. He had originally held that NASA's budget should not be allowed to climb over half a million dollars. Yet by the time he left office, NASA was employing sixteen thousand employees, was spending nearly $1 billion a year, and had plans for spending much more.[100] One of those plans involved initiating work on a manned lunar expedition. During his last months in office, Eisenhower scored at least one clear victory in his effort to contain the space race when he refused to approve such work. Appalled at PSAC's price tag of $26–38 billion to put a man on the moon, he dismissed the lunar expedition as a "multi-billion-dollar project of no immediate value."[101] In his final budget message to Congress in January 1961, Eisenhower refused to include the funds NASA had requested for post-Mercury space exploration.

Conclusion

As Eisenhower left office, there was a widespread impression that he had moved too slowly in the arena of space exploration and ballistic missile development. The contemporary image was that of a president who was tired and uncreative. Eisenhower was seen as failing to grasp both the quickening pace of technological development and the intense anxiety that Americans felt about falling behind in this area.

The far greater resources that Eisenhower's successor, John F. Kennedy, committed to space seemed further to confirm the charge that Eisenhower's response to Soviet gains in space had been inadequate. The younger, vibrant Kennedy, it appeared, understood what the aging Eisenhower had not—that the America couldn't afford to lose the space race and that bold steps were needed to rehabilitate its self-image of technological prowess. During the 1960s and early 1970s, as the space age unfolded and Project Apollo succeeded, Eisenhower's legacy for space policy appeared all the more questionable. If the U.S. space program had continued at the moderate pace established by Eisenhower, America might never have made it to the Moon. Yet, as with his presidency in general, Eisenhower's record on space appears different with the benefit of hindsight and the extensive declassification of documents on his presidency.

The argument of this chapter has been that early U.S. space policy reflected elements of Eisenhower's political philosophy and leadership style that have come to be more clearly recognized and appreciated by scholars in recent years. Following *Sputnik,* Eisenhower projected calm during a time of near panic, patiently explaining to the public why American security was not at risk. In an atmosphere of intense cold war competition, he resisted conceptualizing space exploration as an out-and-out race with the Soviets and sought, albeit with mixed success, and used the power of his office to place the pursuit of U.S. space abilities within a balanced program for boosting American scientific prowess. During the last three years of his presidency, Eisenhower resisted enormous political pressures to launch a crash U.S. defense effort. Nonetheless, he failed to convince the American public of the validity of his position, and in that sense his leadership was less successful than he had intended. Within just ten months of *Sputnik*'s launch, Eisenhower had been forced to create NASA to carry out space exploration as an independent civilian agency.

Eisenhower was not a visionary when it came to space policy. Instead, while he was intrigued with the idea of space, he remained a consistent skeptic about the entire enterprise of exploring the heavens. This outlook reflected his larger strategic framework, not a passive approach to the presidency or a failure of imagination. Despite the confusing nature of the space issue, Eisenhower had a clear idea from the mid-1950s onward of what type of space program he wanted. Through sustained engagement in space policy, especially after *Sputnik,* he used the power of the rising "imperial presidency" to implement the kind of cautious program that he believed was most appropriate for the time.

It is never easy for former presidents to watch their successors shift the direction of national policy. During his post–White House years, Eisenhower (who resumed the rank of general of the Army) ordinarily resisted criticizing his successors on matters bearing on national security. Still, he was dismayed at President Kennedy's announcement in May 1961 that the United States would place a man on the Moon by the end of the decade. This decision dramatically reversed one that Eisenhower had made just six months earlier. In Eisenhower's view, this decision appeared not only ill advised but clearly motivated by political expediency—namely, the desire for the Kennedy administration to regain its momentum after the failed Bay of Pigs invasion of April 1961.

In a 1965 letter to Major Frank Borman, a NASA astronaut who had been troubled by Eisenhower's criticism of Project Apollo, Eisenhower explained his thinking: "What I have criticized about the current space program is the concept under which it was drastically revised and expanded just after the Bay of Pigs fiasco." Eisenhower wrote that he thought a race to the Moon was unwise and that it distorted America's space program. It "immediately took one single project or experiment out of a thoughtfully planned and continuing program involving communication, meteorology, reconnaissance, and future military and scientific benefits and gave the highest priority—unfortunate in my opinion—to a race, in other words, a stunt."[102]

For the most part, Eisenhower kept quiet about his views on Project Apollo. He did not mount a public campaign against the undertaking or devote whole speeches and articles to criticizing Kennedy's space policy. His most pointed public criticism came in an August 1962 *Saturday Evening Post* article that dealt with a wide range of public issues. "By all means, we must carry on our explorations in space," Eisenhower wrote, "but frankly I do not see the need for continuing this effort as such a fantastically expensive crash program." Eisenhower expressed his dismay that NASA was requesting $4 billion a year for space and that this budget figure was headed further upward. He said he felt as proud as anyone about the successes of U.S. astronauts. "But why the great hurry to get to the moon and planets? We have already demonstrated that in everything except the power of our booster rockets we are leading the world in scientific space exploration. From here on, I think we should proceed in an orderly, scientific way, building one accomplishment on another, rather than engaging in a mad effort to win a stunt race."

In the same article, Eisenhower reiterated his long-held views on prestige: "If we must compete with Soviet Russia for world 'prestige,' why not channel the struggle more along the lines in which we excel . . . ? Let's put some other items in this 'prestige' race: our unique industrial accomplishments, our cars for almost everybody instead of just the favored few, our remarkable agricultural productivity, [and] our supermarkets loaded with a profusion of appetizing foods." Eisenhower's central point, which he could not stress enough while president, was that the cold war competition had many fronts and the United States should fight on those where it was strongest.[103]

It was this approach to the cold war that most distinguished Eisenhower from his successor. In contrast to Eisenhower, Kennedy held that the struggle with the Soviet Union had to be waged in every category of power and in every part of the world. He viewed the psychological component of the struggle to be centrally important, for, as he had so often emphasized as a senator, much of the developing world was still ideologically uncommitted and could be lost to communism if the United States stumbled. In the realm of defense policy, Kennedy argued that it was not enough to rely on the blunt threat of massive retaliation. Instead, the United States must be able to fight and win on each rung in the ladder of escalation, from guerrilla insurgency to conventional war, and even nuclear exchanges. Just as crucially, U.S. willingness to fight had to be totally credible. What all this meant was that America's prestige—the perception abroad of its overall strength and vitality—could not be in question if the United States were to remain secure. In space policy, this thinking underpinned a strong determination to beat decisively the Soviets in what Kennedy, unlike Eisenhower, readily acknowledged was a race. Kennedy used his presidential power to carry out this effort, and his announcement of the Apollo decision was one aspect of his response to the cold war.

The question of which philosophy of cold war competition was best suited to the 1950s and 1960s is so dependent on subjective judgment as to be unanswerable in any final way. Still, a number of conclusions can be drawn. First, as the Kennedy administration reluctantly acknowledged shortly after taking office, there was no missile gap. Rather, the United States was far stronger than the Soviet Union in the area of missiles and the larger area of strategic potency.[104] Hence, the claim that lack of accomplishment in space was a sign of military weakness was not valid.

Second, the Soviet success in beating the United States into space with the *Sputnik* launch did not materially damage the Western cold war position. There were no capitulations to communism by borderline countries or diplomatic concessions to Moscow made by the United States and its NATO allies. In particular, Khrushchev's post-*Sputnik* missile rattling in Europe won him no tangible gains. If anything, the Soviet triumph in space served to enhance both American security and prosperity by galvanizing the United States to devote additional resources to education and technological innovation.[105]

Finally, and more generally, it is evident that the low priority Eisenhower placed on prestige in the cold war did not result in any international setbacks during the 1950s. In contrast, it is possible to link Kennedy's strong emphasis on prestige (and that of his advisors who went on to serve under Johnson) with America's fiasco in Vietnam. Eight years after a newly inaugurated Kennedy put forth the view that no front in the cold war could be ignored, America successfully landed men on the Moon and returned them safely to earth in one of the greatest technological feats of the twentieth century. But in that same year, 1969, America began its retreat from Vietnam—the greatest foreign policy disaster in U.S. history. Arguably, both the triumph of Project Apollo and the calamity of the Vietnam War were outgrowths of the same national security philosophy.

With the passage of time, Eisenhower's broad conception of national prestige has come to be more widely appreciated. In the late 1950s, when America dominated the global economy, Eisenhower seemed old-fashioned and lacking in economic sophistication when he insisted that American prosperity could not be taken for granted and that budgetary irresponsibility could threaten that prosperity. At the end of the twentieth century, these sentiments do not seem so misplaced. Likewise, Eisenhower was clearly ahead of his time when he stressed that America's economic performance and its standard of living were as important, if not more important, to U.S. prestige than military might and space exploits.

Notes

1. Eisenhower, "President's News Conference of February 3, 1960," *Public Papers of the Presidents of the United States: Dwight D. Eisenhower, 1960–61* (Washington, D.C.: Government Printing Office, 1964), pp. 144–45.

2. On the transformation of Eisenhower's reputation from that of a passive and mediocre president to that of an informed, involved policy maker (whatever the merits of his policies), see Steven Rabe, "Eisenhower Revisionism: A Decade of Scholarship," *Diplomatic History* 17 (Winter 1993): 97–115. For a comprehensive review of the literature on Eisenhower's presidency, including the recent scholarship that accounts for the remarkable improvement of his historical reputation, see the bibliographical essay in Chester J. Pach Jr. and Elmo Richardson, *The Presidency of Dwight D. Eisenhower* (Lawrence: University of Kansas Press, 1991), pp. 263–72. For an analysis of Eisenhower's leadership style, see Fred I. Greenstein, *The Hidden-Hand Presidency: Eisenhower as Leader* (New York: Basic Books, 1982).

3. See, in particular, John Lewis Gaddis, *Strategies of Containment: A Critical Appraisal of Postwar American National Security Policy* (New York: Oxford University Press, 1983); Richard H. Immerman, "Confessions of an Eisenhower Revisionist," *Diplomatic History* 14 (Summer 1990): 319–42.

4. The most thorough treatment of space policy under Eisenhower can found be in Walter A. McDougall, . . . *The Heavens and the Earth: A Political History of the Space Age* (New York: Basic Books, 1985). For a carefully documented account of the Eisenhower administration's response to the Soviet launching of *Sputnik*, see Robert A. Divine, *The Sputnik Challenge: Eisenhower's Response to the Soviet Satellite* (New York: Oxford University Press, 1993). For an excellent earlier examination of the formation of space policy under Eisenhower, see Enid Curtis Bok Schoettle, "The Establishment of NASA," in *Knowledge and Power: Essays on Science and Government*, ed. Sanford A. Lakoff (New York: Free Press, 1966), pp. 162–70. See also Rip Bulkeley, *The Sputniks Crisis and Early United States Space Policy: A Critique of the Historiography of Space* (Bloomington: Indiana University Press, 1991). For studies of the missile-gap controversy, see Desmond

Ball, *Politics and Force Levels: The Strategic Missile Program of the Kennedy Administration* (Berkeley: University of California Press, 1980); Edgar M. Bottome, *The Missile Gap: A Study of the Formulation of Military and Political Policy* (Rutherford, N.J.: Fairleigh Dickinson University Press, 1971); Peter J. Roman, *Eisenhower and the Missile Gap* (Ithaca, N.Y.: Cornell University Press, 1995).

5. Immerman, "Confessions of an Eisenhower Revisionist," p. 328.

6. For two recent examinations of Eisenhower and economic policy, see Ivan W. Morgan, *Eisenhower versus the Spenders* (New York: St. Martin's Press, 1990); John W. Sloan, *Eisenhower and the Management of Prosperity* (Lawrence: University of Kansas Press, 1991). See also James L. Sundquist, *Politics and Policy: The Eisenhower, Kennedy, and Johnson Years* (Washington, D.C.: Brookings Institution, 1968).

7. Quoted by Gaddis, *Strategies of Containment*, p. 134.

8. Morgan, *Eisenhower versus the Spenders*, p. 211.

9. Ibid., p. 51.

10. Dwight D. Eisenhower, *The White House Years: Waging Peace* (Garden City, N.Y.: Doubleday, 1965), p. 217 (hereinafter cited as *Waging Peace*).

11. Eisenhower, "Farewell Radio and Television Address to the American People," Jan. 17, 1961, *Public Papers of the Presidents of the United States: Dwight D. Eisenhower, 1960–61* (Washington, D.C.: Government Printing Office, 1964), p. 1038.

12. Morgan, *Eisenhower versus the Spenders*, pp. 11, 82. For more on Humphrey's views, see Sloan, *Eisenhower and the Management of Prosperity*, pp. 20–25.

13. Gaddis, *Strategies of Containment*, p. 134.

14. Immerman, "Confessions of an Eisenhower Revisionist," p. 332.

15. Ibid., p. 333.

16. James P. Killian Jr., interview, Mar. 14, 1970, Eisenhower Administration Project, Oral History Research Office, Columbia University, p. 70.

17. Immerman, "Confessions of an Eisenhower Revisionist," p. 335.

18. Dulles, "Address to the Council on Foreign Relations, January 12, 1954," *Department of State Bulletin*, no. 30 (Jan. 25, 1954), p. 108.

19. Gaddis, *Strategies of Containment*, p. 164.

20. Harkey Reiter to Bryce Harlow, n.d. [ca. Oct. 1957], with attachments, Records of Bryce N. Harlow, Dwight D. Eisenhower Library (DDEL), Abilene, Kans.

21. "Report on the Present Status of the Satellite Problem," Aug. 23, 1952, Grosse File, NASA Historical Reference Collection, NASA History Office, Washington, D.C.

22. In deemphasizing prestige, Eisenhower may have been influenced on this issue by U.S. government studies during the 1950s that showed that American prestige overseas was very high and quite solid. See, for example: *Current Trends in Attitudes toward the U.S. and U.S.S.R., West European Public Opinion Barometer Report*, no. 17 (Washington, D.C.: U.S. Information Agency, 1956).

23. McDougall, . . . *The Heavens and the Earth,* p. 119.

24. National Security Council, NSC 5520, *Foreign Relations of the United States, 1955–1957,* 27 vols. (Washington, D.C.: Government Printing Office, 1988), 11:723–30 (hereinafter cited as FRUS).

25. Ibid., p. 730.

26. Ibid., pp. 734–42.

27. Eisenhower, *Waging Peace,* p. 209.

28. James R. Killian Jr., *Sputnik, Scientists, and Eisenhower: A Memoir of the First Special Assistant to the President for Science and Technology* (Cambridge, Mass.: MIT Press, 1977), p. 76.

29. Eisenhower, *Waging Peace,* p. 208.

30. Killian, *Sputnik, Scientists, and Eisenhower,* p. 119.

31. Eisenhower, "Summary of Important Facts in the Development by the United States of an Earth Satellite," Oct. 9, 1957, NASA Historical Reference Collection.

32. Quoted in William E. Burrows, *Deep Black: Space Espionage and National Security* (New York: Random House, 1987), p. 67.

33. For background on the U-2 program and Eisenhower's feelings about it, see Michael R. Beschloss, *Mayday: Eisenhower, Khrushchev, and the U-2 Affair* (New York: Harper and Row, 1986).

34. McDougall, . . . *The Heavens and the Earth,* p. 123.

35. FRUS, 11:749.

36. Working Group on Certain Aspects of NSC 5520, memorandum of meeting, June 13, 1957, NASA Papers, [1], Working Group on Certain Aspects of the Scientific Earth Satellite—Minutes 1956–58, DDEL.

37. For a comprehensive assessment of Eisenhower's response to *Sputnik,* see Divine, *Sputnik Challenge.*

38. Bulkeley, *Sputniks Crisis,* p. 5.

39. McDougall, . . . *The Heavens and the Earth,* p. 153.

40. Burrows, *Deep Black,* p. 90.

41. Eisenhower, *Waging Peace,* p. 205.

42. Ibid., p. 206.

43. Killian, *Sputnik, Scientists, and Eisenhower,* p. 7.

44. Ibid., p. 10.

45. Eisenhower, *Waging Peace,* p. 211.

46. *Facts on File* 17, no. 884: 330.

47. Ibid., p. 331.

48. Eisenhower, *Waging Peace,* p. 224.

49. Eisenhower, "Summary of the Important Facts."

50. Divine, *Sputnik Challenge,* p. 10; Roger D. Launius, *NASA: A History of the U.S. Civil Space Program* (Malabar, Fla.: Krieger Publishing, 1994), pp. 22–23.

51. National Security Council, minutes, Oct. 10, 1957, Ann Whitman File, NSC series, box 9, DDEL.

52. Legislative Leadership Meeting, supp. notes, Jan. 7, 1958, Legislative Meetings series, box 3, DDEL..

53. Eisenhower to Loyd S. Swenson Jr., Aug. 5, 1965, NASA Historical Reference Collection.

54. See for example, Memorandums of Conference with the President, Feb. 17, 1959, and Oct. 21, 1959, box 12, Records of the White House Office of Science and Technology (OST).

55. Ibid.

56. Discussion at the 431st meeting of the National Security Council, Jan. 12, 1960, Ann Whitman File, NSC series, box 12, DDEL.

57. FRUS, 24:162–63.

58. Ibid., p. 164.

59. FRUS, 24:183–84.

60. Senate Special Committee on Space and Astronautics, *Compilation of Materials on Space and Astronautics,* 85th Cong., 2d sess. (Washington, D.C.: Government Printing Office, 1958), p. 1.

61. McDougall, . . . *The Heavens and the Earth,* p. 178.

62. Ibid., p. 171.

63. National Security Council, NSC 5814/1, "Preliminary U.S. Policy on Outer Space," Aug. 18, 1958, Office of the Special Assistant for National Security Affairs, box 67, DDEL.

64. See National Security Council, NSC 5918, "U.S. Policy on Outer Space," Dec. 17, 1959, Office of the Special Assistant for National Security Affairs, box 70, DDEL.

65. "Introduction to Outer Space," White House, Mar. 26, 1958, NASA Historical Reference Collection.

66. T. Keith Glennan, *The Birth of Nasa: The Diary of T. Keith Glennan,* ed. J. D. Hunley (Washington, D.C.: NASA [SP-4105], 1993), p. 2.

67. Ibid., p. 27.

68. Quoted in Killian, *Sputnik, Scientists, and Eisenhower,* p. 143.

69. Gen. Andrew J. Goodpaster Jr., interview, July 22, 1974, p. 56, NASA Historical Reference Collection.

70. For an in-depth analysis of the missile gap controversy, see Bottome, *Missile Gap.*

71. Some have speculated that Eisenhower's refusal to condemn the Democrats for the late start in space research and development may have been more pragmatic than indicated here. In essence, Eisenhower might have been hesitant to politicize the issue because he could also have been condemned by the Democrats and stood as much to lose as to gain. Unfortunately, we do not know his motivations and can only conclude that while he certainly had a strong case to point fingers at the Truman administration—and collected information during the *Sputnik* crisis to be able to do so—he chose not to, probably for a mixture of reasons ranging from altruistic to pragmatic.

72. Eisenhower, *Waging Peace*, pp. 207–10, quotation is from p. 207.

73. Greenstein, *Hidden-Hand Presidency*, pp. 57–99.

74. Divine, *Sputnik Challenge*, p. 45.

75. Ibid., p. 76.

76. Bulkeley, *Sputniks Crisis*, pp. 128–29.

77. Killian, *Sputnik, Scientists, and Eisenhower*, pp. 221, 228.

78. Fred Kaplan, *The Wizards of Armageddon* (New York: Simon and Schuster, 1983), p. 136.

79. National Security Council, NSC 5724, "Deterrence and Survival in the Nuclear Age," Security Resources Panel of the Science Advisory Committee, Nov. 7, 1957, NASA Historical Reference Collection.

80. Gaddis, *Strategies of Containment*, p. 359.

81. Memorandum of Conference with the President, Feb. 6, 1958, box 12, OST.

82. Eisenhower, *Waging Peace*, 257.

83. For a detailed look at the fight over control of space policy, see McDougall, . . . *The Heavens and the Earth*, pp. 157–176; Schoettle, "Establishment of NASA," pp. 162–269.

84. Killian, *Sputnik, Scientists, and Eisenhower*, p. 122.

85. Schoettle, "Establishment of NASA," p. 233.

86. Killian, *Sputnik, Scientists, and Eisenhower*, p. 131.

87. Schoettle, "Establishment of NASA," p. 233.

88. Ibid., p. 192.

89. Killian, *Sputnik, Scientists, and Eisenhower*, p. 228.

90. Ibid., p. 126.

91. Memorandum of Conference with the President, Feb. 6, 1958.

92. Killian, *Sputnik, Scientists, and Eisenhower*, p. 122.

93. Glennan, *First Years*, pp. 7–8.

94. Eisenhower, *Waging Peace*, p. 260.

95. National Security Office, NSC 5814/1, "Preliminary U.S. Policy on Outer Space."

96. Memorandum of Conference with the President, Feb. 24, 1959, box, 12, OST.

97. McDougall, . . . *The Heavens and the Earth*, p. 200.

98. John M. Logsdon, *The Decision to Go to the Moon: Project Apollo and the National Interest* (Cambridge, Mass.: MIT Press, 1970), p. 35.

99. Glennan, *First Years*, p. 18.

100. Jane Van Nimmen and Leonard C. Bruno, with Robert L. Rosholt, *NASA Historical Data Book*, vol. 1, *NASA Resources, 1958–1968* (Washington, D.C.: NASA [SP-4012], 1988), pp. 70, 137–38. The Eisenhower administration had requested $915 million for FY 1961 with a supplemental request on Jan. 18, 1961, for an additional $49.6 million. The Eisenhower budget request for FY 1962 was $1.1 billion. Of the roughly 16,000 employees in NASA in early 1961, almost 8,000 had come from the NACA, and many others had transferred from the military Vanguard and Saturn programs.

101. Stephen E. Ambrose, *Eisenhower,* vol. 2, *The President* (New York: Simon and Schuster, 1984), p. 591; Logsdon, *Decision to Go to the Moon,* pp. 35–36.

102. Ambrose, *Eisenhower,* p. 641.

103. Eisenhower, "Are We Headed in the Wrong Direction?" *Saturday Evening Post,* Aug. 11, 1962, p. 24.

104. Kaplan, *Wizards of Armageddon,* p. 295.

105. See, in particular, Barbara Barksdale Clowse, *Brainpower for the Cold War: The Sputnik Crisis and the National Defense Education Act of 1958* (Westport, Conn.: Greenwood Press, 1981).

2

Kennedy and the Decision to Go to the Moon

Michael R. Beschloss

In his 1960 presidential campaign, John F. Kennedy never called explicitly for a crash effort to put an American on the Moon by 1970, but his campaign rhetoric pointed in the direction of greater activism in space. Kennedy's critique of Dwight D. Eisenhower and Richard M. Nixon centered around the charge that the incumbent administration had placed the United States in danger of slipping behind the Soviet Union in the cold war. Kennedy pledged, if elected, to make the United States a nation that was not "first but, first and, first when, first if, but first PERIOD."[1]

That desire, as well as Kennedy's faith in the power of science and technology to accomplish great feats, led to the 1961 decision to go to the Moon. Kennedy used the growing power of the "imperial presidency"— derived in part from the cold war effort to empower experts, in this case aerospace engineers, with the responsibility and wherewithal to execute a "crash" program—to place Americans first on the Moon.[2]

Kennedy and Khrushchev

Kennedy framed his desire for American leadership in terms of military and economic strength as well as international prestige. American shortcomings in space gave him a powerful symbol in all three areas. In his effort to demonstrate American inferiority, Kennedy thus exploited during his presidency the issue of space exploration in exactly the same way that Nikita Khrushchev was doing and to the same effect. In the military field, Kennedy was not averse to leaving Americans somewhat in the dark about the distinction between advances in space exploration and advances in production of ICBMs.

Like Khrushchev, who correctly gambled that launching *Sputnik* would lead many people in the world to conclude that the Soviet Union had suddenly gained an important form of military superiority, Kennedy hammered Eisenhower and Nixon for failing to keep up with the Soviets in rocket thrust development. He hoped that this would strengthen his charge that the United States was suffering from a "missile gap," lagging behind the Soviets in ICBMs. During the 1960 debates, Kennedy told Nixon, "You yourself said to Khrushchev, 'You may be ahead of us in rocket thrust, but we're ahead of you in color television.'"[3] Elsewhere Kennedy said, "I will take my television in black and white. I want to be ahead of them in rocket thrust."[4]

Khrushchev had argued that Soviet space achievements were an emblem and dividend of superior Soviet economic growth rates. So had Kennedy. Khrushchev had exploited his space triumphs to suggest to newly emerging developing nations that his was the best political and economic system to emulate. Kennedy too argued that American failures in space weakened U.S. prestige, and he produced a series of U.S. Information Agency poll findings to prove it. With amazing overstatement—even for a campaign—Kennedy insisted in New York in October 1960, "The key decision which [the Eisenhower] administration had to make in the field of international policy and prestige and power and influence was their recognition of the significance of outer space. . . . The Soviet Union is now first in outer space."[5]

Thus, when Kennedy was elected in November 1960, he was compelled to use the power of his office to make dramatic gestures, as he put it, to "turn the tide" back in favor of the United States in the cold war.[6] This was difficult to do with the "missile gap." If he had any doubt before the election, he had none afterward when given U.S. classified information that the missile gap had been a false issue and that the United States held a large lead over the Soviet Union in ICBMs.

It was also difficult to turn the tide with economic growth. If Kennedy had any doubt during the campaign, he knew after he took office that his comparison of superior Soviet growth rates to those of the United States had been bogus. Not only was he privy to classified information that demonstrated the desperate weaknesses of the Soviet economy, he also knew that the reason why Soviet economic growth looked so much better than America's was because the 1959 figures he had used were taken in the middle of the worst U.S. recession in years, that the Soviet rate was artificially inflated by cheating, and that the Soviet economy was recovering from the devastation of World War II. All of this increased Kennedy's motivation in December 1960 to find some quick way of seeming to boost the American position in the cold war and vindicate the rhetoric of his campaign.

The Definition of a Space Policy

At that moment, Eisenhower as outgoing president received a classified report of an ad hoc panel on manned spaceflight. The panel asked "whether the presence of a man adds to the variety or quality of the observations which can be made from unmanned vehicles—in short, whether there is a scientific justification to include man in space vehicles." Its answer was a polite "no": "Man's senses can be satisfactorily duplicated at remote locations by the use of available instrumentation. . . . It seems, therefore, to us at the present time that man-in-space cannot be justified on purely scientific grounds. . . . On the other hand, it may be argued that much of the motivation and drive for the scientific exploration of space is derived from the dream of man's getting into space himself."[7]

This finding dovetailed perfectly with Eisenhower's views. If anything, Eisenhower had a tin ear for the effect of space achievements on America's international position. In 1957, he had not dreamed that the launching of an earth satellite could have had remotely the impact that *Sputnik* did on Soviet prestige—and he refused to be stampeded afterward by senators such as Lyndon B. Johnson who demanded that the United States catch up. Eisenhower felt that spending on space exploration could be seriously defended only in military and scientific terms. He felt that among the various forms of space exploration, manned spaceflight should be nothing more than one instrument in the symphony. He was unwilling to "hock my jewels" to support the enormous cost of sending an American quickly to the Moon, which he regarded as a "stunt." He "couldn't care less whether a man ever reached the moon."[8] And he used his presidential power to circumvent other politicians' plans to increase space activities that he deemed unwise. As a result, the civilian space effort in the Eisenhower administration was moderate and measured, much to the chagrin of its advocates.

Moreover, Eisenhower, in setting up the National Aeronautics and Space Administration (NASA), had used his presidential power to put in charge of it people who shared his perspective on space exploration and how aggressively it should be pursued. NASA's first administrator, T. Keith Glennan, perfectly reflected Eisenhower's priorities in space. Glennan emphasized a well-rounded, measured space program that did not focus on "spectacular" missions designed to "one up" the Soviets. He also believed that the new space agency should remain relatively small and that much of its work would of necessity be done under contract to private industry and educational institutions.[9] Hugh L. Dryden, Glennan's deputy, repeatedly expressed caution about competition with the Soviets in any space race. On April 16, 1958, for example, Dryden testified before a House committee that a Defense Department human spaceflight proposal had "about the same technical value as the circus stunt of shooting the young lady from the gun" and lacked any scientific merit.[10]

None of this pleased Kennedy. He was disappointed in January 1961, when as president-elect he received the report of his own task force on space, chaired by Jerome Wiesner, who was to become his White House science advisor. Interestingly, virtually every member of the panel had been deeply involved as outside consultants in the Eisenhower administration's space policy making. Thus, the Wiesner Report, which was written for public consumption, was not the ringing denunciation of Eisenhower's lassitude on space initiatives that Kennedy and his entourage might have hoped for.

The report conceded that "during the next few years, the prestige of the United States will in part be determined by the leadership we demonstrate in space activities"—and that recent U.S. accomplishments in space had "not been impressive enough." Still, as far as manned exploration was concerned, it was "very unlikely that we shall be the first in placing a man into orbit around the earth." The panel warned that "space activities are so unbelievably expensive and people working in this field are so imaginative that the space program could easily grow to cost many more billions of dollars per year."[11]

The Wiesner Report's conclusion was not too different from Eisenhower's: human spaceflight should not be given an exaggerated amount of attention in the context of other space activities. This did not delight the new president. Kennedy treated the panel's findings like a skunk at a wedding. He told reporters, "I don't think anyone is suggesting that their views are necessarily in every case the right views."[12] As with so many areas of his policy toward the Soviet Union during the first two months of his presidency, Kennedy played for time on space and kept his options open.

In March 1961, sensing an opportunity, the new NASA administrator, James E. Webb, asked Kennedy's budget director, David Bell, for a 30 percent increase in the budget his agency had been allocated by Eisenhower. Bell wrote to Kennedy that he wondered whether the United States should run races it might lose anyway, that there were other better and cheaper ways of enhancing American prestige, and that "the total magnitude of present and projected expenditures in the space area may be way out of line with the real values of the benefits."[13] Bell told Dryden of NASA that he had better be patient, because the president had other problems to worry about. Dryden replied, with some prescience if little feeling, "You may not feel he has the time, but whether he likes it or not, he is going to have to consider it. Events will force this."[14]

That same week, Kennedy met with Webb, who on taking office was eager "to make unmistakably clear our support for the manned spaceflight program."[15] Webb had been recommended to Kennedy by his business associate, the oil man Robert Kerr, Democrat from Oklahoma, who was also the new chairman of the Senate's Space Committee. Webb, a denizen of Washington political circles since the 1930s who had served as Truman's budget director and as an undersecretary of state, had fled the city only with Eisenhower's election in 1953. Now Webb was back, eager to use the power of the federal government to accomplish "New Deal–type" programs on behalf of the nation—nor was he the kind of man to tolerate an America that was "first if" or "first when." He would not be content with a modest mission or budget for the agency he had made some financial sacrifices to oversee.[16]

Using language that played to his presidential audience, Webb told Kennedy:

> The extent to which we are leaders in space science and technology will in large measure determine the extent to which we, as a nation, pioneering on a new frontier, will be in a position to develop the emerging world forces and make it the basis for new concepts and applications in education, communications and transportation, looking toward more viable political, social and economic systems for nations willing to work with us in the years ahead.[17]

Webb made a sale. In his defense message to Congress on March 25, 1961, Kennedy asked for $125.7 million for the kind of large boosters that would lead to a Moon mission. Still he was marking time. He told Webb that he would not make any final decision on the main elements of the NASA request until the fall of 1961.[18]

Crisis

Kennedy's plans were abruptly changed by two unexpected events in mid-April. On April 12, for the first time, the Soviets launched a man, Yuri Gagarin, into Earth orbit, creating a worldwide space sensation dwarfed only by *Sputnik I.* One NASA scientist summarized the perspective of space exploration advocates: "Wait until the Russians send up three men, then six, then a laboratory, start hooking them together and then send back a few pictures of New York for us to see."[19]

On the day after the most recent Soviet triumph, Webb went to the Oval Office. Like Eisenhower after *Sputnik,* he was not spooked by Gagarin's flight. Webb said, "The solid, onward, step-by-step pace of our program is what we are more interested in than being first."[20] To boost Kennedy's spirits, Webb carried a desk model of the Mercury capsule that would soon take the first American into space. Kennedy had enough of a sense of humor to tell one aide afterward that Webb had probably bought it in a toy store that morning. Nonetheless, he asked NASA for a study of the feasibility and costs of an accelerated civilian space program.

Kennedy could easily afford to tolerate the Gagarin success. After less than three months in office, he knew that he could not be blamed for the American disadvantage he had criticized so sharply on the campaign trail. Then, one week later, the sky fell when CIA-backed Cuban exiles failed in their invasion of Fidel Castro's Cuba at the Bay of Pigs.

No matter how much Kennedy's aides tried, through background interviews with reporters, to shift the blame for the Bay of Pigs fiasco onto Eisenhower—and they did—Kennedy knew that the debacle had the power to shatter his entire administration. The Bay of Pigs suggested to Americans that they had elected a president who was at the least inexperienced and at worst incompetent. Especially after American reversals in Laos and the Congo, Kennedy's failure in Cuba was exactly the kind of cold war setback that he had denounced throughout the campaign and pledged to avoid if he were elected president. He was desperately in need of something that would divert the attention of the public and identify him with a cause that would unify them behind his administration.

On April 20, the day Kennedy knew for certain that the Bay of Pigs had failed, he called in Vice President Johnson, chairman of his Space Council, and asked him to come up with something fast in space. He gave Johnson a memo that was redolent of presidential panic: "Do we have a chance of beating the Soviets by putting a laboratory in space, or by a trip around the moon, or by a rocket to land on the moon, or by a rocket to go to the moon and back with a man? Is there any other space program which promises dramatic results in which we could win?" Kennedy went on, "How much additional would it cost? Are we working 24 hours a day on existing programs? If not, why not? . . . Are we making maximum effort? Are we achieving necessary results? . . . I would appreciate a report on this at the earliest possible moment."[21] That day, Kennedy told reporters, "If we can get to the moon before the Russians, we should."[22]

By giving the chairmanship of his Space Council to Johnson, who as a Senator had used *Sputnik* to good political advantage in criticizing the Eisenhower administration for not investing more in a space program, Kennedy had tipped the scales in the direction of an aggressive effort in space. While in the Senate, Johnson had if anything been more extreme than Kennedy in his demands for an accelerated space effort. After *Sputnik* he had grandiloquently exclaimed that the nation that controlled the "high ground" of outer space had the capacity to rule the world.[23]

Consensus Building

Johnson turned to NASA for information to answer the president's questions on what to do in space to "beat" the Soviets. On April 22, 1961, NASA's Dryden responded to the request about a Moon program by writing that there was "a chance for the U.S. to be the first to land a man on the moon and return him to earth if a determined national effort is made." He added that the earliest this feat could be accomplished was 1967, but that to do so would cost about $33 billion, a figure $10 billion more than the entire projected NASA budget for the next ten years.[24]

Johnson also asked Secretary of Defense Robert S. McNamara for his views. McNamara knew that Kennedy was moving toward a review of space policy and, after three months in the New Frontier, he was already adept at the kind of language and arguments that would win the favor of this president. With the exception of Attorney General Robert Kennedy, McNamara had already proven himself the dominant figure in the Kennedy cabinet. Aside from knowing that Kennedy wanted an accelerated space program, McNamara had another motivation: the increased effort would make a perfect customer for companies in the aerospace industry that were already irate over the cutbacks McNamara was planning in the U.S. defense program. McNamara flatly wrote to Johnson, "Major achievements in space contribute to national prestige. This is true even though the scientific, commercial or military value of the undertaking may, by ordinary standards, be marginal or economically unjustified. What the Soviets do and what they are likely to do are therefore matters of great importance from the viewpoint of national prestige."[25]

Johnson also canvassed friends in private business, as well other government officials, including the fabled space scientist Wernher von Braun, who had built V-2s for the Nazis in World War II and come to the United States in 1945. Von Braun told him that the United States had "a sporting chance of sending a 3–man crew *around the moon* ahead of the Soviets" and "an excellent chance of beating the Soviets to the *first landing of a crew on the moon* (including return capability, of course)." He added,

> The reason is that a performance jump by a factor of ten over their present rockets is necessary to accomplish this feat. While today we do not have such a rocket, it is unlikely that the Soviets have it. Therefore we would not have to enter the race toward this obvious next goal in space exploration against hopeless odds favoring the Soviets. With an all-out crash program I think we could accomplish this objective in 1967–1968.

Von Braun ominously concluded, "I do not believe that we can win this race unless we take at least some measures which thus far have been considered acceptable only in times of a national emergency."[26]

After garnering these technical opinions, Johnson began to persuade political leaders of the need for an aggressive lunar landing program. He brought together Senators Kerr and Styles Bridges (R-NH) and spoke with several representatives to ascertain if they were willing to support an accelerated space program. Whenever he heard reservations, Johnson used his forceful personality to persuade. "Now," he asked, "would you rather have us be a second-rate nation or should we spend a little money?" He also persuaded Secretary of State Dean Rusk, a member of the Space Council, to support the initiative. Rusk wrote to the Senate Space Committee, saying, "We must respond to their conditions; otherwise we risk a basic misunderstanding on the part of the uncommitted countries, the Soviet Union, and possibly our allies concerning the direction in which power is moving and where long-term advantage lies." It was clear early in these deliberations that Johnson favored an expanded space program and a maximum effort to land an astronaut on the Moon.[27]

Kennedy's mandate to Johnson had been framed so bluntly and specifically that the vice president was unlikely to return to the Oval Office and tell his boss that he should stop worrying about space and turn to other matters. This was especially true because, in the spring of 1961, Johnson was working hard to maximize his influence on the Kennedy administration. He also knew that if he had any presidential ambitions for 1968, they would depend largely on Kennedy's attitude toward his vice president.

Thus, not surprisingly, on April 28, Johnson gave Kennedy a report that was largely what the president wished to hear. Sounding like Kennedy on the campaign trail, Johnson emphasized:

> The U.S. has greater resources than the U.S.S.R. for attaining space leadership but has failed to make the necessary hard decisions and to marshal those re-

sources to achieve such leadership. . . . This country should be realistic and recognize that other nations . . . will tend to align themselves with the country which they believe will be the world leader—the winner in the long run. Dramatic accomplishments in space are being increasingly identified as a major indicator of world leadership. . . . We are neither making maximum effort nor achieving results necessary if this country is to reach a position of leadership.

Johnson insisted that manned exploration of the Moon was essential, whether or not the United States turned out to be first. In this exercise, Johnson had built, as Kennedy had wanted, a strong justification for a presidential initiative to undertake Project Apollo, but he had also moved further, toward a greater consensus for the objective among key government and business leaders.[28]

While NASA's leaders were enthusiastic about the course Johnson recommended, they wanted to shape it as much as possible to the agency's long-run priorities. NASA's administrator, Webb, well known as a skilled political operator who could seize an opportunity, organized a short-term effort to accelerate and expand a long-range NASA master plan for space exploration. A fundamental part of this effort addressed a legitimate concern that the scientific and technological advancements for which NASA had been created not be eclipsed by the political necessities of international rivalries. Webb conveyed the concern of the agency's technical and scientific community to Wiesner on May 2, 1961, noting that "the most careful consideration must be given to the scientific and technological components of the total program and how to present the picture to the world and to our own nation of a program that has real value and validity and from which solid additions to knowledge can be made, even if every one of the specific so-called 'spectacular' flights or events are done after they have been accomplished by the Russians." He asked that Wiesner help him "make sure that this component of solid, and yet imaginative, total scientific and technological value is built in."[29]

Although the White House agreed that the program should be balanced, with an accelerated Moon landing as its centerpiece, Webb was not yet convinced. He did not wish to undertake a Moon project unless assured that NASA would have full funding and support. Thus, he refused to argue on NASA's behalf for a Moon program. On May 3, Johnson called him to a meeting that included Senator Kerr to suggest that Webb would get what he wanted—and thus get Webb to change his mind. In his notes for the meeting, Johnson wrote, "We are here to discuss not WHETHER, but HOW—not WHEN, but NOW." In vintage Johnsonian language, he compared the space program to his success in bringing electricity to the hill country of Texas. He told Webb, "So far NASA has gotten everything it has asked for. I want them to plan and dream big enough to get us out ahead."[30]

Webb caved in. Five days later, he gave Johnson what he wanted—a letter, written jointly with McNamara, asking for a tacit new doctrine in U.S. space policy that would lead to an Apollo Moon landing before 1970. No longer would the U.S. government follow the principle defined by Eisenhower that projects that were part of the space competition with the Soviet Union had to have other elements of "intrinsic merit." The letter said:

> This nation needs to make a positive decision to pursue space projects aimed at enhancing national prestige. . . . The non-military, non-commercial, non-scientific but "civilian" projects such as lunar and planetary exploration are, in this sense, part of the battle along the fluid front of the cold war. Such undertakings may affect our military strength only indirectly, if at all, but they have an increasing effect upon our national posture. . . . We recommend that our National Space Plan include the objective of manned lunar exploration before the end of this decade. . . . The orbiting of machines is not the same as the orbiting or landing of man. It is man, not merely machines, in space, that captures the imagination of the world. . . . Even if the Soviets get there first, as they may . . . it is better for us to get there second than not at all. . . . If we fail to accept this challenge, it may be interpreted as a lack of national vigor and capacity to respond.[31]

With Johnson on a presidential mission to Southeast Asia, Kennedy discussed the Webb-McNamara letter with his cabinet on May 10. Bell was concerned about setting specific dates for a Moon landing and about spending so much money on prestige. Secretary of Labor Arthur Goldberg disagreed with the premise that a Moon program would stimulate the economy. But Kennedy's intentions were clear. Wiesner later recalled that when McNamara noted that without Apollo there would be a dangerous oversupply of manpower in the aerospace industry, "this took away all argument against the space program."[32]

Decision

At the end of May, Kennedy was to fly to Europe for a summit with Khrushchev. He did not wish to go merely in the wake of American failures in Laos, the Congo, Cuba, and in space. He decided to break presidential tradition by delivering a second State of the Union address on May 25 that would deal with "urgent national needs" and in which he planned to invoke the power of the presidency to initiate an aggressive lunar landing program. That speech asked for the most open-ended commitment ever made in peacetime in order to land an American on the Moon and represented a high moment of the imperial presidency.

As the speech was written, Kennedy argued with his advisors over what date should be announced as the target for the Moon landing. Webb suggested that a late 1968 Moon trip would be a triumphant climax to his second term as president. White House aides more cautiously suggested simply "before this decade is out." They reasoned that this could be interpreted to include 1970.[33] They may also have felt that if no landing occurred before the end of 1970, blame for the failure could be shifted from Kennedy to his successor.

Speaking before Congress, Kennedy tried to avoid the sense that his demand was being hastily made in the wake of Gagarin and the Bay of Pigs. He specifically noted that he had been reviewing U.S. space policy "since early in my term." Space, he said, "may hold the key to our future on earth. . . . I believe that this nation should commit itself to achieving the goal, before this decade is out, of landing a man on the moon and returning him safely to the earth." Kennedy departed from his prepared text—the only time he ever did so before Congress as president—to say, "Unless we are prepared to do the work and bear the burdens to make it successful, there would be no sense in going ahead."[34]

His aide Theodore Sorensen thought that the president's voice sounded "urgent but a little uncertain." Afterward, while riding with Sorensen back to the White House, Kennedy observed that the routine applause that greeted his announcement had sounded "something less than enthusiastic." He said that $20 billion was "a lot of money." The congressmen knew "a lot of better ways to spend it."[35]

Former President Eisenhower wrote to a friend that Kennedy's decision to back a crash program for a Moon landing was "almost hysterical" and "a bit immature."[36] In 1965, he complained to astronaut Frank Borman of how the Moon program "was drastically revised and expanded just after the Bay of Pigs fiasco. . . . It immediately took one single project or experiment out of a thoroughly planned and continuing program involving communication, meteorology, reconnaissance, and future military and scientific benefits and gave the highest priority—unfortunate in my opinion—to a race, in other words, a stunt."[37] But by a nearly unanimous vote Congress agreed with Kennedy, in part because of the intense consensus building by Johnson and other politicians. Project Apollo became the dominant component of the U.S. space program. The U.S. budget for space was increased by 50 percent in 1961. The next year, it exceeded all pre-1961 space budgets combined.

The reaction of Eisenhower's NASA administrator, T. Keith Glennan, was especially insightful of the conservative response to Kennedy's decision. For instance, Glennan told Eisenhower, then in retirement at Gettysburg, Pennsylvania, that "this is a very bad move—that we are entering into a competition which will be exceedingly costly and which will take up an increasingly large share of that small portion of the nation's budget which might be called controllable."[38] Glennan harped on this issue for years, never quite able to understand the Kennedy administration's philosophy that large expenditures for science and technology in the form of a race against the Soviets to the Moon could produce important benefits.

Glennan also told Webb, Kennedy's NASA head, of his dismay at the Apollo mandate:

> I have no doubt at all as to the desirability and inevitability of manned flight to the moon. And I would accept—not willingly—a national decision to beat the Russians to the moon if such a decision resulted in a truly "crash" program with no effort spared or held back. No one knows the intentions of the Soviet Union but all of us understand the ability they have to dedicate men and facilities and treasure to that particular effort they believe desirable or necessary. To enter a "race" against an adversary under such conditions and to state that no additional taxes are necessary—indeed to suggest tax reductions—does not seem to me to be facing facts nor to be completely frank about the on-going program. . . .
>
> There can be only one real reason for such a "race." That reason must be "prestige." The present program without such a "race" but with full intention of accomplishing whatever needs to be accomplished in lunar and planetary exploration, unmanned and/or manned, is a vigorous and costly one. It will produce most of the significant technology and essentially all of the scientific knowledge that will be produced under the impetus of the "race" and at the lower cost in men and money. . . .
>
> No, Jim, I cannot bring myself to believe that we will gain lasting "prestige" by a shot we may make six to eight years from now. I don't think we should play the game according to the rules laid down by our adversary.[39]

The best way to establish Kennedy's personal impact on the decision to go to the Moon is to imagine what might have occurred in the winter and spring of 1961 had Eisenhower been somehow elected to a third term: unmotivated to use space as a battlefield in the cold war, unstampeded by setbacks in Asia, Africa, and Latin America, worried about the rising impact of the military-industrial complex and its academic counterparts, determined to achieve a balanced budget, Eisenhower would no doubt have been content not to have an American astronaut reach the Moon by 1970 and have used the power of his office to resist other initiatives to conduct an accelerated space effort.

It is a measure of Kennedy's aversion to long-term planning and his tendency to be rattled by momentary crises that one may conclude that in the absence of the Gagarin triumph and the Bay of Pigs fiasco in April 1961, he might never have gone to the length of asking Congress to spend $20 billion on a crash Moon program. Kennedy's desire for a quick, theatrical reversal of his new administration's flagging position, especially just before a summit with Khrushchev, is a more potent explanation of his Apollo decision than any other. Johnson's desire for turf, McNamara's desire to use aerospace overcapacity, Kennedy's own conviction that a Moon program was consistent with what Sorensen called "the New Frontier spirit of discovery"[40]—these things helped the decision along, but none was so important.

Assessment

Kennedy's commitment to space captured the American imagination and attracted overwhelming support. No high official at the time seemed deeply concerned about either the difficulty or the expense. Congressional debate was perfunctory, and NASA actually found itself literally pressed to spend the funds committed to it for the early 1960s. Kennedy may have understood that the lunar landing was so far beyond the capabilities of either the United States or the Soviet Union in 1961 that the early lead in space activities taken by the Soviets would not predetermine the outcome. As a result, the situation gave Americans a reasonable chance of overtaking the Soviets in space activities and recovering a measure of lost status. Even so, Kennedy's political objectives were essentially achieved by the presidential decision to go to the Moon, and he did not necessarily think much about the long-term consequences.

As Kennedy conceded, his decision for an accelerated Moon landing was ultimately a political decision made in terms of cold war strategy. How does it stand up now that the cold war is over? Not well. We now know that the reason the Soviet Union gave up in that struggle was that it recognized that it could not compete with Western economies and Western societies in those areas of life and death that mattered. Although the Moon program contributed a great deal to the United States, the tens of billions of dollars spent in the 1960s on what Kennedy essentially thought of as world propaganda could probably have been better devoted to U.S. defense or the American domestic economy, and that might have convinced the Soviets more quickly of the fruitlessness of the tragic conflict with the United States.

As taxpayers complained about the cost and scientists about the slighting of more important projects, Republicans began using the word "moondoggle" and "science fiction stunt."[41] In 1962, Kennedy was shown hints that the Soviets were not going to compete with the United States for the first Moon landing. By April 1963, he was asking Johnson for advice on how the Apollo Program could be justified in terms other than cold war prestige. Johnson replied with the reassuring old argument that "our space program has an overriding urgency that cannot be calculated solely in terms of industrial, scientific or military development. The future of society is at stake."[42]

In the fall of 1963 at the United Nations, Kennedy made his most serious public insistence that the United States and Soviet Union explore the Moon together. We shall never know for certain whether this was predominantly an effort, in the wake of the Partial Nuclear Test Ban Treaty, to relax the cold war, or an effort by Kennedy to back away gracefully from an expensive Moon race from which the other side also seemed to be withdrawing.

Notes

1. Michael R. Beschloss, *The Crisis Years: Kennedy and Khrushchev, 1960–1963* (New York: Harper, 1991), p. 28.

2. This deference to the authority of expertise was also seen in other technical arenas. See Bruce E. Seely, *Building the American Highway System: Engineers as Policy Makers* (Philadelphia: Temple University Press, 1987); Samuel P. Hays, with Barbara D. Hays, *Beauty, Health, and Permanence: Environmental Politics in the United States, 1955–1985* (Cambridge: Cambridge University Press, 1987); Thomas L. Haskell, ed., *The Authority of Experts: Studies in History and Theory* (Bloomington: Indiana University Press, 1984); John G. Gunnell, "The Technocratic Image and the Theory of Technocracy," *Technology and Culture* 23 (July 1982): 392–416; Mark H. Rose and Bruce E. Seely, "Getting the Interstate System Built: Road Engineers and the Implementation of Public Policy, 1955–1985," *Journal of Public Policy* 2 (1990): 23–55.

3. Senate, *Joint Appearances of Senator John F. Kennedy and Vice President Richard M. Nixon,* 86th Cong., 2d sess. (Washington, D.C.: Government Printing Office, 1961), p. 211.

4. Senate, *The Speeches of Senator John F. Kennedy: Presidential Campaign of 1960,* 87th Cong., 1st sess. (Washington, D.C.: Government Printing Office, 1961), p. 113.

5. Ibid.

6. Kennedy, "Annual Message to the Congress on the State of the Union," Jan.

30, 1961, *Public Papers of the Presidents of the United States: John F. Kennedy, 1961* (Washington, D.C.: Government Printing Office, 1962), pp. 22–27.

7. President's Science Advisory Committee, "Report of the Ad Hoc Panel on Man-in-Space," Dec. 16, 1960, in *Exploring the Unknown: Selected Documents in the History of the U.S. Civil Space Program,* vol. 1: *Organizing for Exploration,* ed. John M. Logsdon, with Linda J. Lear, Jannelle Warren-Findley, Ray A. Williamson, and Dwayne A. Day (Washington, D.C.: NASA [SP-4218], 1995), pp. 408–12.

8. T. Keith Glennan, *The Birth of NASA: The Diary of T. Keith Glennan,* ed. J. D. Hunley (Washington, D.C.: NASA [SP-4105], 1993), pp. 292–93.

9. These themes are well developed in Glennan's diary. See also, "Glennan Announces First Details of the New Space Agency Organization," Oct. 5, 1958, NASA Historical Reference Collection, NASA History Office, Washington, D.C.; James R. Killian Jr., *Sputnik, Scientists, and Eisenhower: A Memoir of the First Special Assistant to the President for Science and Technology* (Cambridge, Mass.: MIT Press, 1977), pp. 141–44; Killian, interview, July 23, 1974, NASA Historical Reference Collection. Eisenhower's concerns about this aspect of modern America are revealed in "Farewell Radio and Television Address to the American People," Jan. 17, 1961, *Public Papers of the Presidents of the United States: Dwight D. Eisenhower, 1960–61* (Washington, D.C.: Government Printing Office, 1962), pp. 1035–40.

10. House Select Committee on Astronautics and Space Exploration, *Astronautics and Space Exploration, Hearings on H.R. 11881,* 85th Cong., 2d sess. (Washington, D.C.: Government Printing Office, 1958), p. 117.

11. "Report to the President-Elect of the Ad Hoc Committee on Space," Jan. 10, 1961, in *Exploring the Unknown,* ed. Logsdon et al., pp. 416–23.

12. "Space Report," *Washington Post,* Jan. 26, 1961, clipping in "NASA Current News folder," Jan. 1961, NASA Historical Reference Collection.

13. Bell to Kennedy, "NASA Budget Problem," [Mar. 1961], NASA Historical Reference Collection.

14. Quoted in John M. Logsdon, *The Decision to Go to the Moon: Project Apollo and the National Interest* (Cambridge, Mass.: MIT Press, 1970), p. 91.

15. Walter A. McDougall, . . . *The Heavens and the Earth: A Political History of the Space Age* (New York: Basic Books, 1985), p. 312.

16. This is well documented in W. Henry Lambright, *Powering Apollo: James E. Webb of NASA* (Baltimore: Johns Hopkins University Press, 1995).

17. James E. Webb, "Administrator's Presentation to the President," Mar. 21, 1961, NASA Historical Reference Collection.

18. McDougall, . . . *The Heavens and the Earth,* pp. 317–18.

19. Quoted in Hugh Sidey, *John F. Kennedy, President* (New York: Atheneum, 1964), p. 99.

20. Quoted in Logsdon, *Decision to Go to the Moon,* p. 105.

21. Kennedy to Johnson, memorandum, Apr. 20, 1961, in *Exploring the Unknown,* ed. Logsdon et al., pp. 423–24.

22. "U.S. Studying Entry in Race for Moon," *New York Times,* Apr. 22, 1961, clipping in "NASA Current News folder," Apr. 1961, NASA Historical Reference Collection.

23. "52 Million Order Is Placed by Army to Speed Missiles," *New York Times,* Jan. 8, 1958, clipping in "NASA Current News folder," 1957–Apr. 1958, NASA Historical Reference Collection.

24. Dryden to Johnson, Apr. 22, 1961, Vice Presidential Security File, box 17, John F. Kennedy Library, Boston; Logsdon, *Decision to Go to the Moon,* pp. 59–61, 112–14.

25. McNamara to Johnson, "Brief Analysis of Department of Defense Space Program Efforts," Apr. 21, 1961, NASA Historical Reference Collection.

26. Von Braun to Johnson, Apr. 29, 1961, NASA Historical Reference Collection.

27. Robert A. Divine, "Lyndon B. Johnson and the Politics of Space," in *The Johnson Years,* vol. 2, *Vietnam, the Environment, and Science,* ed. Robert A. Divine (Lawrence: University Press of Kansas, 1987), pp. 231–33.

28. Johnson to Kennedy, memorandum, Apr. 28, 1961, NASA Historical Reference Collection.

29. In reality, some of Webb's concern may have been window dressing. Throughout the Apollo years, NASA pushed many projects of arguably greater scientific merit to the back burner and did little long-range planning, sacrificing those concerns for the overtly political goal of beating the Soviets. Many in NASA, of course, had a vested interest in the broader scientific and technical goals of Project Apollo, and Webb tried to maximize them as much as possible within the political environment in which he worked, but he fully recognized that the Kennedy administration was committed to a cold war victory in space almost to the exclusion of any other benefits that might accrue. If there were any other benefits, he knew they would come later.

30. Johnson, opening statement and notes, Vice President's Ad Hoc Meeting, May 3, 1961, NASA Historical Reference Collection.

31. Webb and McNamara to Kennedy, May 8, 1961, in *Exploring the Unknown,* ed. Logsdon et al., pp. 439–52.

32. Jerome B. Wiesner, interview, July 24, 1974, NASA Historical Reference Collection.

33. Theodore C. Sorensen, *Kennedy* (New York: Harper, 1965), p. 525.

34. Kennedy, "Presidential Address on Urgent National Needs," reading text, May 25, 1961, in *Exploring the Unknown,* ed. Logsdon et al., pp. 453–54.

35. Sorensen, *Kennedy,* p. 526.

36. Beschloss, *Crisis Years,* p. 166.

37. Ibid.

38. Glennan to Eisenhower, May 31, 1961, Glennan Personal Papers, 19DD4, Archives, Case Western Reserve University, Cleveland, Ohio.

39. Glennan to Webb, July 21, 1961, Glennan Personal Papers, 19DD4.

40. Sorensen, *Kennedy,* p. 525.

41. See McDougall, . . . *The Heavens and the Earth,* pp. 389–93.

42. Kennedy to Johnson, Apr. 9, 1963, and Johnson to Kennedy, May 13, 1963, NASA Historical Reference Collection.

3

Johnson, Project Apollo, and the Politics of Space Program Planning

Robert Dallek

Lyndon Johnson was a difficult, imperious character with a penchant for overheated rhetoric and big political plans. He left a record of landmark social gains and disastrous public failures, always using his presidential office to the hilt. Civil rights, voting rights, and Medicare alone are enough to give him a place in twentieth-century American history with Franklin Roosevelt, the greatest domestic reform president in the national experience. Johnson's spectacular failure in Vietnam would suffice to label him as one of the worst foreign policy leaders in the country's history. In the nearly thirty years since he left the White House we have not come to terms with this political giant. Indeed, this generation of Americans probably never will. Memories of Johnson's many transgressions against the national self-esteem remain too fresh to allow a sufficiently detached assessment of the man's impact on the country's life. One hopes this will change in time. For we need to see Johnson's career not as a chance to indulge our sense of moral superiority but as an opportunity to gain an understanding of many subjects crucial to the nation's past and future.

Space policy seems as good a place as any to begin. The major part Johnson played in shaping the country's space program in the 1950s and 1960s did not provoke then, nor does it now, the kind of controversy we associate with his war on poverty, the Great Society, and the Vietnam War. Moreover, Johnson's views of space tell us a great deal about his whole political career: that is, about his priorities and the means he used to achieve them. More specifically, four considerations determined Johnson's thinking about space policy in 1957–69: national security, personal political and party gain, domestic social advance, and budgetary constraints. None of these concerns, however, operated to the exclusion of the others. To be sure, at one time or the other each of these goals became the dominant motive in determining LBJ's response to changing circumstances at home and abroad, but the other aims were never far from his mind. Yet however much Johnson's motives altered over time in dealing with space matters and however much his levels of support for space exploration rose and fell, especially in the last years of his presidency, he deserves to be remembered as the elected official who did as much, if not more, for space exploration than any other American political leader in this century.

Johnson and Early Space Policy

Sputnik I and *II,* the Soviet Earth satellites launched in October and November 1957, spurred Johnson's initial interest in fostering an aggressive American space program. His first aim was to advance the country's missile technology and eliminate a "missile gap" between the United States and the Soviet Union. Second, he believed that promoting a space program was good politics for himself and his party. "The issue [*Sputnik*] is one which, if properly handled, would blast the Republicans out of the water, unify the Democratic party, and elect you President," George Reedy, a principal Senate aide, told him. "I think you should plan to plunge heavily into this one." Johnson saw the political advantage to himself and the Democratic party in seizing the space issue. But he feared a witch-hunt that might undermine confidence in the country's military strength and encourage the belief that we could not meet the Soviet challenge.

Johnson genuinely put the national security issue first in trying to design a response to the Soviet's demonstrated superiority in the space race. During the winter of 1957–58, as chairman of an Armed Services subcommittee on preparedness, he held hearings on how the United States could produce better missiles at a faster rate. The hearings' sole objective, he declared, was securing the defense of the United States; he had no interest in finger-pointing or assessing blame for past mistakes and wished to use the past strictly as a guide for future action. John Steele, *Time*'s congressional correspondent, told his editors that Johnson would "run a good investigation" that would serve a useful purpose. There would be no "political witch-hunt. Johnson knows that a good investigation is the only kind that will satisfy anyone, and in the end bring credit to anyone. . . . Here, as downtown [at the White House], there is a sense of urgency, of consideration of the national interest."

Yet Johnson was not simply a selfless patriot. As one official at the Department of Defense (DOD) said, "No sooner had *Sputnik*'s first beep-beep been heard—via the press—than the nation's legislators leaped forward like heavy drinkers hearing a cork pop." The facts emerging from Johnson's investigation demonstrated the Eisenhower administration's ineptness in mounting an effective missile and space program. It also allowed Johnson to identify himself as the country's leading congressional advocate of a stepped-up effort in space. He dominated the hearings, introducing witnesses, leading cross-examinations, and making himself the principal spokesman to the press. In January 1958, he told the Senate Democratic caucus that "control of space means control of the world" and urged his party colleagues to sign on to a greatly expanded space effort. Later that month, at the conclusion of the hearings, he persuaded his subcommittee to issue seventeen recommendations that, without being overtly partisan, showed Johnson and the Democrats as pushing the Eisenhower administration into what they thought was essential for the national well-being.

The journalists Rowland Evans and Robert Novak described Johnson's handling of the *Sputnik* crisis as "a minor masterpiece." Without involving himself in a direct collision with Eisenhower, Johnson used the space issue to damage the White House and benefit himself and the Democrats. Yet at the same time, he served the nation by propelling it into the space age. Specifically, he took the leading role in Congress in sponsoring legislation to create a National Aeronautics and Space Administration (NASA). Although Johnson's aides did much of the work on the space bill, he played a significant part in shaping NASA's organization. While understanding that the military would have a large say in any space program, *he argued successfully* for making NASA a civilian agency. Such a move would avoid service rivalries and satisfy political demands for peaceful uses of space. "The space program was a paramilitary operation in the cold war, no matter who ran it," the historian Walter A. McDougall wrote, but civilian control headed off a significant imbalance between the services and met the political needs of American officials at home and abroad.[1]

The Vice President as Space Czar

Johnson's election to the vice presidency in 1960 gave him a continuing role in space policy. This defied the tradition of consigning a vice president to the outer fringes of power. The office of vice president, Thomas R. Marshall, Woodrow Wilson's vice president observed, "is like a man in a cataleptic state. He cannot speak. He cannot move. He suffers no pain. And yet—he is conscious of all that goes on around him." "The chief embarrassment in discussing his [the vice president's] office," Wilson wrote, "is that in explaining how little there is to be said about it, one has evidently said all there is to say." Johnson, who had a lifelong aversion to being anything but top dog, later described the vice presidency as "nothing," saying "I detested every minute of it." Daniel Patrick Moynihan remembered looking into Vice President Johnson's eyes and thinking, "This is a bull castrated very late in life."[2]

Though Johnson's vice presidency was largely a ceremonial job, his role in space matters went beyond what a vice president normally did. In 1961, President Kennedy persuaded Congress to amend the 1958 space law to make the vice president, instead of the president, the chairman of the Space Council, an advisory group that President Eisenhower *had largely ignored* in 1958–61. Kennedy had no intention of letting Johnson eclipse him on a matter given high public visibility by Soviet space shots, but he was eager to use Johnson's expertise on something of vital national concern. Moreover, in giving Johnson some prominence as an architect of America's space program, Kennedy was making him a political lightning rod. Should an effort to catch and pass the Soviets in space technology fail or suffer a well-publicized defeat, Johnson would be out front taking some, if not much, of the heat.[3]

Yet Johnson eagerly accepted the risk. He saw American achievements in space as vital to the cold war contest with the Soviet Union. The Soviets' more advanced space program in 1957–61 persuaded Kennedy, Johnson, and millions of Americans that they were falling behind not only in missile technology but also in the global competition for "hearts and minds."

Consequently, in April 1961, after a Soviet cosmonaut became the first man to orbit Earth and the failure at the Bay of Pigs had embarrassed the United States, Kennedy asked Johnson to make "an overall survey of where we stand in space. Do we have a chance of beating the Soviets by putting a laboratory in space, or by a trip around the moon, or by a rocket to land on the moon, or by a rocket to go to the moon and back with a man? Is there any other space program which promises dramatic results in which we could win?" Johnson replied that the Soviets were ahead of us "in world prestige attained through technological accomplishments in space." And other nations, identifying space gains as a reflection of world leadership, were being drawn to the Soviets. A strong effort was needed at once to catch and surpass the Soviets if the United States were to win "control over . . . men's minds through space accomplishments." Johnson recommended "manned exploration of the moon" as "an achievement with great propaganda value." "The real 'competition' in outer space," he said, was between the communist and free enterprise social systems. The control of outer space was going to "determine which system of society and government [would] dominate the future. . . . In the eyes of the world, first in space means first, period; second in space is second in everything." When people complained about the cost of space exploration, Johnson replied: "Now, would you rather have us be a second-rate nation or should we spend a little money?"[4]

Kennedy needed no prodding from Johnson to make the case for some dramatic space venture. At the end of May 1961, he told a joint session of Congress:

> If we are to win the battle that is now going on around the world between freedom and tyranny, the dramatic achievements in space which occurred in recent weeks [a suborbital flight by astronaut Alan Shepard] should have made clear to us all . . . the impact of this adventure on the minds of men everywhere, who are trying to make a determination on which road they should take. . . . Now it is . . . time for this nation to take a clearly leading role in space achievement, which in many ways may hold the key to our future on earth.

Kennedy asked the country to commit itself to the goal of landing an American on the Moon and returning him safely to Earth before the decade was out.[5]

Yet Kennedy worried that a highly publicized American space effort that ended in failure would further damage the nation's prestige and inflict a political wound that could jeopardize his hold on the presidency. Shepard's flight had encouraged Kennedy's hopes that America might catch and pass the Soviets, but he remained concerned about future mishaps. In June, when Shepard drove with President Kennedy, Johnson, and Newton Minow, head of the Federal Communications Commission, to speak before the National Convention of Broadcasters, Kennedy poked Johnson and said: "You know, Lyndon, nobody knows that the Vice President is the Chairman of the Space Council. But if that flight had been a flop, I guarantee you that everybody would have known that you were the Chairman." Everyone laughed, except Lyndon, who looked glum and angry, especially after Minow chimed in, "Mr. President, if the flight would have been a flop, the Vice President would have been the next astronaut."[6]

The possibility that he would be a political sacrificial lamb for a faulty space effort did not dampen Johnson's enthusiasm for a manned mission to the Moon. His commitment rested partly on his faith in liberal nationalism, the ability of government to ensure economic and social progress through the use of its largesse. For Johnson, whose whole career had been built on the assumption that federal monies, well spent on infrastructure, social programs, and defense, could serve the national well-being, but especially in the less affluent South, the space program was a splendid way to serve the country's defense, expand the domestic economy, and advance scientific understanding. In 1963, when criticism from academics, journalists, and political conservatives began to be heard against "the moondoggle," Johnson told Kennedy, "The space program is expensive, but it can be justified as a solid investment which will give ample returns in security, prestige, knowledge, and material benefits." During a plane trip as vice president to visit various space installations around the United States, Johnson gave an hour-long "very impassioned talk" to Minow on the virtues of communications satellites in both advancing education in underdeveloped countries and educational television in the United States.[7]

Johnson also saw other, more selfish benefits flowing from the space program. Convinced he was backing a winner, he made strong efforts to identify himself with every aspect of its work. Not only did he crisscross the country in publicized visits to space installations, he also gave a series of "factual space reports to the public" on the work of NASA and his Space Council. The ostensible objective was to educate the country, but it had the added advantage of keeping his name in the news.[8]

Then there were the pork-barrel gains that served the economic interests of Texas and the South and strengthened his political hold on the state and the region, especially at a time when his support of civil rights for African Americans was undermining it. Although Johnson denied any part in the selection of southwestern companies receiving Apollo or Moon program contracts or in shifting half of the space operations from Cape Canaveral in Florida to a command center in Houston, Senator George Smathers (D-FL) knew better. "He and I had a big argument about it, big fight," Smathers remembered. "Johnson tried to act like he didn't know. . . . It never has made sense to have a big operation at Cape Canaveral and another big operation in Texas. But that's what we got, and we got that because Kennedy allowed Johnson to become the theoretical head of the space program." Indeed, with Robert Kerr of Oklahoma, a Johnson friend, running the Senate Space Committee; Texas Congressmen Overton Brooks and Olin Teague the House counterparts; Albert Thomas, another Texas representative, chairing the Appropriations Committee; and James Webb, Johnson's nominee, directing NASA, the southwest generally and Texas in particular profited most from Kennedy's accelerated space program.[9]

In 1962, when lobbyists and congressmen from outside the South began to complain about a southwest monopoly on NASA contracts, Kennedy made Richard L. Callaghan, a congressional staffer, an assistant administrator to James Webb. Callaghan's job was to arrange for a more equitable distribution of contracts, which would relieve congressional pressure on Kenny O'Donnell, the president's liaison to Congress, and find out whether Kerr and Johnson were pulling strings at NASA for their friends. As Callaghan later told Robert Sherrod, a *Time-Life* reporter: "'Kenny O'Donnell wasn't only interested in getting the contractors off his back. He wanted to satisfy himself about the Kerr-Johnson influence on the Space Agency. He wanted to find out who was getting what—wanted to satisfy himself that the organization was honest.' VERY INTERESTING," Sherrod wrote in a note to himself. "OBVIOUSLY JFK PUT O'DONNELL UP TO PLANTING CALLAGHAN, BUT HOW TO PROVE THAT THE PRESIDENT WAS SUSPICIOUS OF THE BIGGEST TEXAS WHEELER DEALER OF THEM ALL, AND OF THE 'KING OF THE SENATE,' WHO SUCCEEDED HIM AS CHAIRMAN OF THE SPACE COMMITTEE?" According to what Sherrod later learned from O'Donnell, there was no evidence to prove any wrongdoing by anyone at NASA. Nor could they find anything on Johnson that might have made him a potential liability to the Kennedy administration. As Johnson himself later ruefully noted, "The damn press always accused me of things I didn't do. They never once found out about the things I did do."[10]

Johnson's thousand days as vice president justifiably enhanced his reputation as someone who saw substantial national benefits flowing from expanded U.S. space efforts. It also demonstrated his effectiveness in building a national consensus for a space program. As Webb later told a BBC interviewer,

> When President Kennedy asked him [LBJ] to prepare a memorandum as to what our space programme should be, . . . he called in some businessmen. . . . Then he called in Wernher von Braun and General Schriever from the Air Force and a large number of technical people and sort of had hearings. As we approached the end of that, he called in the political leaders . . . in Congress and he in effect said to them: "We ought to go forward but we don't want to go forward unless you are going to commit yourself to stay with us." . . . So he developed this commitment of certain leaders . . . and this you see made it a lot easier for the rest of the country to come along. They saw that these very powerful, responsible people, both political people in the Congress and business people from outside, believed this should be done, then we will accept it and go forward.[11]

A Space Advocate in the White House

During his first year as president, from November 1963 to November 1964, Johnson pushed hard to keep the space effort on track. Although deter-

mined to keep his first budget under $100 billion in order to win passage
of Kennedy's $11 billion tax cut pending before Congress, Johnson agreed
to increase NASA spending by $150 million to $5.25 billion. "Our plan
to place a man on the moon in this decade remains unchanged," he told
Congress in January 1964. "It is an ambitious and important goal. In
addition to providing great scientific benefits, it will demonstrate that our
capability in space is second to no other nation's." But, he emphasized,
"we cannot reach this goal without sufficient funds. There is no second-
class ticket to space."[12]

At the same time, Johnson's decision to press ahead with Project Apollo—the U.S. Moon landing—rested less now than in 1961–63 on considerations of national security. In May 1963, he had declared, "I do not think
this generation of Americans is willing to go to bed each night by the light
of a Communist moon." During the first year of his presidency, he remained
eager to beat the Soviets in the space race, but a U.S. missile buildup under Kennedy, JFK's success in the Cuban missile crisis in 1962, and the
Nuclear Test Ban Treaty in 1963 had eased concerns about a missile gap
and fears that we had fallen behind the Soviets in military might and scientific research. On the day of his assassination, Kennedy himself had intended to say "that there was no longer any fear that a Communist lead
in space would become the basis of military superiority."[13]

Some worries about these matters remained, but during the first half of
1964 Johnson put greater emphasis on working out cooperative agreements
with the Soviet Union to explore outer space. "President Johnson has apparently lost his enthusiasm for the Soviet-American space race," the *New
York Herald Tribune* reported in June 1964. Earlier in the year, the president had sent the deputy administrator of NASA, Hugh L. Dryden, to
Geneva "to seek agreements for a 'widening area' of cooperation in space
with Moscow." Judging from National Security action memoranda in
1964, Johnson was clearly eager for less competition and more cooperation with the Soviets in space. As the astronaut and later Senator John
Glenn saw it, the Congress was no longer so easily moved to increase space
spending by appeals to the Soviet threat. "The anti-Russian theme had
worn out," Glenn remembered. Johnson, ever sensitive to congressional
moods, saw the need to press the case for space exploration on other
grounds.[14]

A more compelling consideration with Johnson, especially at the start
of his presidency, was to carry out Kennedy's agenda. Johnson had to con-
front the grief and despair many people felt over the assassination of a
beloved leader and their antagonism toward someone who, however much

he might identify himself with JFK, seemed like a usurper, an unelected, untested replacement for the man the country now more than ever saw as more suitable for the job. In the first days of his presidency, only 5 percent of the public felt they knew very much about Johnson, while 67 percent said they knew next to nothing about him. Seventy percent had doubts about how the country would "carry on without" Kennedy. Seeing an essential need for continuity, for reassurance that the new president would be faithful to the previous administration's ends and means, Johnson made fulfillment of Kennedy's promise to put a man on the Moon and safely return him to Earth by 1970 one of his major priorities.[15]

Apollo: A Great Society Initiative

Johnson also believed that the Apollo mission made excellent economic and political sense. Landing a man on the Moon would not only reaffirm America's superiority over the Soviet Union and honor Kennedy's memory, it would also spur both immediate and long-term economic growth and gain the administration considerable political credit with the public. Less than a month after becoming president, Johnson was pressing NASA to use its resources to help Wisconsin and Minnesota expand "their research and engineering capabilities." Webb, who was a savvy politician in his own right and understood perfectly the importance of tying NASA to specific economic benefits around the country, laid plans to double NASA's "activity" in both states. More importantly, he kept close track of how NASA affected the nation's economy and took every opportunity to apprise Johnson of these gains. In a 1965 report to the president, for example, he pointed out that in the previous year 94 percent of NASA's "procurement dollars" had gone to 20,000 private U.S. industrial companies: $331 million had been spent in 120 cities in 22 states with high unemployment rates, and as many as 750,000 people worked directly or indirectly on NASA-related business.[16]

Johnson understood that much more than pork-barrel spending would result from NASA's efforts generally and Project Apollo in particular. To be sure, as a seasoned politician with a keen appreciation for federal largesse, he greatly valued the economic and political gains coming to localities and his White House from NASA's spending in Florida, Alabama, Texas, Oklahoma, California, and other states around the country. But he also placed considerable value on the long-term national advances that NASA's work seemed likely to produce. As Webb told him, NASA's accomplishments were leading to the development of "new materials . . . [and] new structures" as well as "complex electronic, mechanical and chemical systems. . . . This new technology . . . is bringing with it revolutionary change in the way of making and testing things, not only for space systems, but for innumerable other non-space services, processes and materials."

Because these benefits were essentially abstractions, Webb took pains to enumerate the many more concrete returns flowing from NASA's research and development. He told Johnson NASA had something to offer to law enforcement in terms of data processing and communication systems; to the construction industry through NASA developed materials; to pollution control through the development of an outlook whereby the Earth's air and water are beginning to be viewed as finite resources operating as closed systems; to transportation of people in and out of the inner city through research on short-haul aircraft; to improvement of economic opportunities for all citizens by stimulating business through new inventions and transfers of space technology to industry; and to a richer life by development of techniques making possible cheaper, lighter, and more reliable television sets and other electronic items for use in the home.[17]

For Johnson, the work of space exploration was part of a "Great Society," a larger vision he enunciated in May 1964. In a speech at the University of Michigan, he appealed to the best in the American temperament:

For a century we labored to settle and to subdue a continent. For half a century we called upon unbounded invention and untiring industry to create an order of plenty for all of our people. The challenge of the next half century is whether we have the wisdom to use that wealth to enrich and elevate our national life, and to advance the quality of our American civilization. . . . For in your time we have the opportunity to move not only toward the rich society and the powerful society, but upward to the Great Society. . . . It is a place where men are more concerned with the quality of their goals than the quantity of their goods.

To reach this promised land, Americans would have to pledge themselves to a crusade for excellence. "For better or for worse, your generation has been appointed by history . . . to lead America toward a new age," he insisted. "Will you join in the battle to build the Great Society, to prove that our material progress is only the foundation on which we will build a richer life of mind and spirit?"[18]

"An obvious component of this [Great Society] theme," one White House aide told Edward Welsh of the Space Council, "is the vast array of implications of our present Research and Development activity." Webb understood perfectly what Johnson had in mind. "I know of no area," he told the president, "where the inspirational thrust toward doing everything required of a great society can be better provided on a proven base of competence, and with so many practical additional benefits to be derived, than through the space program. . . . The space program lies in your first area of building the great society, for it is truly an imaginative new program based on new ideas and new capabilities." Early in 1965, after becoming vice president, Hubert Humphrey echoed Webb's point in a speech at the Goddard Memorial Dinner:

> Let me assure you that the Great Society envisioned by President Johnson is not one limited to the fight against poverty, ignorance, disease, and intolerance. The Great Society requires, in addition, an urgent quest for excellence, for intellectual attainment, for crossing new frontiers in science and technology. Let me emphasize that an adequately funded, well-directed space program is an integral part of our nation's commitment to its future, to its greatness.[19]

Johnson himself told a group of astronauts in 1965 that their missions not only increased "our knowledge of technology" but also would lead "to a better life for all." In a 1969 interview, Johnson said that plans to get to the Moon had inspired the country to do something about its educational systems, medical care for the elderly, conservation, and poverty. In his 1971 memoirs, he recalled: "Space was the platform from which the social revolution of the 1960s was launched. We broke out of far more than the atmosphere with our space program. . . . If we could send a man to the moon, we knew we should be able to send a poor boy to school and to provide decent medical care for the aged. In hundreds of other forms the space program had an impact on our lives." A few of the benefits he saw the country reaping from investments in space included pacemakers for heart patients, intercontinental television, lightweight electronics equipment to improve navigation techniques for ships and planes, more abundant food supplies, improved conservation of natural resources, and weather control capabilities that saved lives, crops, and cattle.[20]

If space exploration tied into Johnson's hopes for a Great Society, it also served his political purposes in the 1964 presidential campaign. Johnson's opponent, the conservative Arizona Senator Barry Goldwater, complained, "We are spending entirely too much money on the manned moon program." He promised that as president he would have "all manned space research . . . directed by the military" and would use the "billions of dollars saved from abandoning the manned lunar program" for "military space missions." As with so many other issues in the campaign, Goldwater was out of sync with the national mood. Polls in the spring and fall of 1964 showed 64–69 percent of the public favorably disposed to landing an American on the Moon, with 78 percent saying the Apollo Program should be maintained at its current pace or speeded up. Only 20 percent of the country supported space spending strictly for military purposes. In response, Johnson refused "to slacken in our nationally approved effort to reach the moon as soon as we can." Identification with the widely backed Apollo mission was superb politics in an election year.[21]

Johnson and the Budget Crisis

While Johnson gave wholehearted support to Apollo, he also thought about what, if any, big projects might come next. In January 1964, he asked Webb to describe NASA's future plans, specifically asking how "hardware and development programs" would be tied to "prospective missions." In May, Webb provided a tentative answer in which he said that NASA had "virtually completed the investment in facilities" that would land astronauts on the Moon and "meet a broad range of not yet specified tasks." These might include a greater mastery of space science, which would improve weather prediction and control; exploration of the Moon to expand our understanding of the origins of the solar system; a search for life on other planets; the development of space stations, manned and unmanned; better weather, communications, and navigation satellites; and exploration of the near planets and probes of more distant ones.

Not until February 1965, however, did Webb give the president a more precise statement of NASA's future plans. Sensing that Johnson, with expanding commitments at home and abroad, was not eager for new big spending on space, Webb backed away from most of the proposals he had identified in his May 1964 letter. Instead, he urged commitments to two more modest programs: the exploration of Mars through an unmanned landing and further exploration of the Moon with the technology developed for Apollo. The distinguishing features of Webb's proposals, an aide told Johnson, were the absence of a request for any "major new launch vehicle systems" and a continuation of NASA funding at current levels.[22]

With Apollo still years away from fulfillment, Johnson was unwilling to make new commitments of any kind. When Webb asked permission to give the chairmen of the House and Senate Space Committees copies of his February letter to inform the Congress about possible future NASA projects, Johnson resisted. "Why do we need to do anything?" he asked in a reply to Jack Valenti, his aide handling the matter. "I would think I would have more leeway & running room by saying nothing[,] which I would prefer."[23]

Beginning in 1965, Johnson took a two-track approach to NASA and space exploration. His only priority was landing a man on the Moon by the end of the decade, as Kennedy and then he had promised. Beyond that, he resisted significant commitments to post-Apollo planning that would cost billions of dollars and engage the country's prestige and energy. One of the striking features of Johnson's memoirs on his presidency is that he devotes only seventeen of six hundred pages to a discussion of space. And of those seventeen pages, only three describe space policy during his presidency. The rest focuses on his Senate and vice presidential years, the period 1957–63, when he felt he had done his most important work for the space program. Indeed, in an interview with Walter Cronkite in 1969, Johnson noted: "Very frankly, I think I spent more time in the space field in '57 and '58 and '59 and '60, and up to '63, than I did after I became President."[24]

This is not to say that Johnson lost interest in space achievements. He attended closely to the various space missions in 1965–68. As Welsh remembered, Johnson watched each mission on television. "He had the astronauts in to see him at the White House. He had them to the ranch. He followed them with a real sense of personal interest. As a matter of fact, he said that he really in a sense flew with them on every flight from the beginning of the launch till they landed safely." Johnson himself told Cronkite: "I have ridden on every mission. . . . I've watched with eagerness, and pride, their every movement."[25]

Nevertheless, his interest did not translate into support for post-Apollo projects. Everything that had initially spurred Johnson to back a major American effort in space—fear of Soviet superiority and a desire for economic and political gains—now became reasons to avoid substantial commitments to new big space programs. Johnson's concern, for example, that Soviet advances in space might undermine America's national security and prestige in the Soviet-U.S. competition for global influence steadily faded from view during his presidential years. In the spring of 1966, after the Soviets had landed an unmanned spaceship on the Moon, Webb pressed the president to use the Soviet feat to extract more money for NASA from Congress. Webb told Johnson that he had done his best to "minimize the political risk to your Administration from the fact that we are operating substantially under what would be the most efficient program." This was Webb's way of warning that the Soviets might beat the United States to a manned Moon landing, for which Johnson would pay a high political price.[26]

But Johnson was not impressed. He was justifiably confident that the United States would land men on the Moon ahead of the Soviets, and he was certain that Moscow was now more eager for cooperation than competition with the United States in space. Indeed, nine days before Webb's warning about the continuing Soviet threat to America's leadership in space, Johnson had issued "a statement outlining the essential elements of a celestial bodies treaty" and asked U.N. Ambassador Arthur Goldberg to initiate discussions. During the next three months, Soviet and American negotiators drafted nine initial articles of an outer space treaty. By December, additional points of agreement were incorporated into the treaty, which Johnson now publicly described as the "most important arms control development since the Limited Test Ban Treaty of 1963." The space treaty, signed in January 1967 and entered into force in October, banned the placing of weapons of mass destruction in orbit, in outer space, or on celestial bodies, established an unconditional commitment to assist and return astronauts who landed in another country, and forbade claims of sovereignty over celestial bodies.[27]

To Johnson and the State Department, the agreement meant a "de-fusing of the space race" and a reduction or even an end to "much of the pressure to race for new and distant goals." Henry Owen of the department's Policy Planning Council anticipated "strong opposition" from NASA and the Space Council to additional cooperation with the Soviets in space, because it would mean less funding of post-Apollo projects. More cooperation with Moscow and less ambitious space plans, Owen told Walt W. Rostow, chairman of the Policy Planning Council, "will save money, which can go to (i) foreign aid, (ii) domestic purposes—thus mitigating the strain of the war in Vietnam." Owen urged Rostow to get into the fight with NASA and to enlist "someone on the domestic side of the White House staff . . . to ensure that someone . . . representing the constituency whose interests are most directly affected, gets into the fight." A State Department paper entitled "Space Goals after the Lunar Landing" argued that by deemphasizing or stretching out "additional costly programs aimed at the moon and beyond, resources may to some extent be released for other objectives—foreign aid, domestic needs, scientific efforts in other areas—which might serve more immediate, higher priority U.S. interests."[28]

Johnson agreed. The increasing costs of the war in Vietnam, which began to escalate rapidly in 1965, and the outlays for the antipoverty and Great Society programs, which also made substantial budgetary demands beginning in 1965, were central considerations in Johnson's resistance to post-Apollo space commitments. In July 1969, at the end of his administration and after the successful Moon landing, Johnson was vague and evasive about post-Apollo plans. "What would you like to see as the next space goal?" an interviewer asked him. "I don't want to be setting goals for those that are responsible for this effort," he said. "I would like to take all that we have done and be sure that we utilize all the knowledge that we have gained up to now, and to follow through to milk the entire Apollo program of every benefit that can come from it." Johnson then ticked off the various ideas others had for post-Apollo planning: space stations, additional Moon shots, studies in space medicine, and unmanned trips to other planets. Personally, he would not say what he favored but hoped that we would continue to have a vigorous space program.[29]

Johnson's remarks were symptomatic of his refusal to make significant, large-scale commitments beyond Apollo in his 1967–69 budgets. His rhetoric masked the battles he and Webb had fought over funding for NASA's future. After suffering a modest cut of about $75 million in 1966, Webb was determined to increase NASA's funding in 1967. But Johnson wouldn't hear of it. Webb's request for $5.3 billion could not withstand a $300 million reduction. In accepting the president's cut, Webb warned against keeping NASA's funding at the current level for another year. "The 1968 budget will be a major turning point with indicated requirements on the order of $6 billion of new obligational authority," he told Johnson in May 1966. By August, however, it was clear that Congress and the president would drop NASA's funding below $5 billion for FY 1967. This would "leave no choice," Webb berated Johnson, "but to accelerate the rate at which we are carrying on the liquidation of some of the capabilities which we have built up." He predicted that options would now be foreclosed and that doubt and uncertainty would demoralize NASA. And, he bluntly declared, "There has not been a single important new space project started since you became President. Under the 1968 guidelines very little looking to the future can be done next year. . . . I cannot avoid the feeling that this is not in the best interests of the country."[30]

Johnson relied on his budget director, Charles Schultze, to counter Webb's assertions. Schultze argued that NASA's funding was entirely adequate to meet the 1969 deadline for a Moon landing and to work toward more distant goals, such as a Mars landing and/or Earth orbital stations. After all, "the space program is not a WPA," Schultze declared. Neither he nor Johnson felt that NASA's budget was skimpy alongside the $2 billion in spending on elementary and secondary education, $1.8 billion on the poverty program, $200 million on water pollution control, and $25 million for high-speed ground transportation. A $5 billion space budget or even a little below that would not "wreck the space program," Schultze contended, nor would it lead to "the liquidation of some of the capabilities we have built up." NASA's funding did not represent "a lack of support for the space program." Moreover, Schultze did not see how in the context of the fighting in Vietnam the administration could afford to meet Webb's request. Johnson agreed with Schultze and convinced Webb to back his decision publicly, though privately the NASA administrator continued to press his case, unsuccessfully requesting an additional $182 million above the $455 million slated for post-Apollo planning.[31]

Johnson saw little political risk in turning aside Webb's demands for more money. By the end of 1966, it was clear to him that NASA and space exploration beyond the Apollo landing had diminished popular appeal. By the summer of 1965, a third of the nation favored cutting the space budget, while only 16 percent wanted to increase it. Over the next three and a half years, support for cutting space spending went up to 40 percent, with those preferring an increase dropping to 14 percent. A poll taken in the summer of 1969 found that 53 percent of the country was opposed to a manned mission to Mars. At the end of 1967, the *New York Times* reported that a poll conducted in six American cities showed that five other public issues held priority over efforts in outer space. Residents of these cities preferred tackling air and water pollution, job training for unskilled workers, national beautification, and poverty before spending additional federal funds on space. The following year *Newsweek* echoed the *Times*'s findings, stating, "The U.S. space program is in decline. The Viet Nam war and the desperate conditions of the nation's poor and its cities—which make space flight seem, in comparison, like an embarrassing national self-indulgence—have combined to drag down a program where the sky was no longer the limit."[32]

In addition, Congress was strongly disposed to reduce NASA's budget. A White House survey of congressional leaders at the end of 1966 revealed pronounced sentiment for keeping Apollo on track but for cutting NASA spending by skimping on post-Apollo outlays. In this context, Johnson's request in January 1967 for a $5 billion NASA budget for FY 1968, including $455 million for post-Apollo programs, was pretty bold.[33]

Yet Johnson's inclination to be generous with NASA and provide for a modest amount of post-Apollo spending could withstand neither a disastrous fire in an Apollo command module in January 1967 nor a growing U.S. budget deficit spurred by the fighting in Vietnam. On January 27, a fire destroyed the *Apollo I* capsule and killed astronauts Roger B. Chafee, Edward H. White III, and Virgil I. Grissom during a test at Cape Kennedy. In addition to the tragic loss of life, the fire undermined national confidence in NASA, which was now accused of carelessness in trying to move Project

Apollo forward too quickly. The fire, Johnson said later, represented "an all-time low" for the space effort. "I grieved [not only] for the men and their families but [also] . . . for the space organization. I felt very sad and sorry for Jim Webb and all of his loyal employees." Senate hearings raised questions about a great many defects in the spacecraft and brought Webb into sharp conflict with three senators, who saw him as whitewashing NASA's failings. The *New York Times,* which was also highly critical of Webb, said that NASA stood for "Never a Straight Answer." Though the hearings were "unpleasant and embarrassing for NASA . . . on the whole," an administrative history of the agency asserts, "they gave NASA a sympathetic forum in which to explain how a tragedy had come about, and show how it would serve to correct deficiencies." NASA's forthrightness in responding to the failings that produced the fire restored a measure of confidence in the agency and prompted the Senate committee to recommend that NASA continue to move the Apollo Program forward to achieve its goal.[34]

A federal budget crisis in the summer of 1967 dealt NASA another blow. A $29 billion deficit brought on by Vietnam spending persuaded Johnson to ask Congress for a 10 percent increase in income taxes. To persuade Congress, Johnson felt compelled to match the tax increase with spending cuts applied to FY 1968 beginning in October 1967. NASA was targeted for $500 million in reductions. Webb objected that, with NASA "just now getting back up to speed after the interruptions and difficulties associated with the accident," it would be "the straw that breaks the camel's back," meaning, "the momentum we have achieved will be lost." For Johnson, there was no choice, except where to apply the cuts in NASA programs. As before, despite recommendations to the contrary, he stuck to keeping Apollo on schedule, agreeing instead to center cuts on post-Apollo applications and the unmanned landing on Mars. Once again, Webb and NASA had to accommodate themselves to a reduced budget, now, $4.59 billion. In spite of everything, Webb was still able to assure Johnson that "the goal of the manned lunar landing in this decade is preserved."[35]

The cuts genuinely troubled Johnson. Whenever there were reductions, he would tell Webb, "Next year I hope to make up for this." Johnson "had almost supreme confidence that at some point he could give us resources again and that we could catch up," Webb recalled. More specifically, in a message to Webb in September 1967, the president asked him to "be sure to make abundantly clear [to a congressional committee] that I do not choose to take one dime from my budget for space appropriations for this year." The "Congress forced me to agree to effect some reductions or lose the tax bill." While Johnson's message was partly a case of political finger-pointing, he was truly uncomfortable reining in NASA or any government program he believed served the national well-being. He loved to quote Speaker Sam Rayburn's adage: "Any jackass can kick a barn down, but it takes a good carpenter to build one." More to the point, his whole polit-ical career had been focused on building and using government programs to expand the economy, raise living standards, relieve privation, and build his Great Society. Overreaching himself by trying to institute domestic reforms and fight a war at the same time, he could not find the means to spend simultaneously on guns and butter. It was a reality he found difficult to accept.[36]

Webb also struggled against the reality of declining commitments to NASA. In November 1967, he pressed Johnson's budget director to urge a strong statement by the president about NASA funding when signing its appropriation bill. NASA's congressional backers, Webb said, saw the president as having "'knifed' the very activities he had previously been urging them to support." Johnson's inconsistent leadership, Webb added, had created "bitter feelings . . . in a number of quarters." Webb wanted Johnson to emphasize the continuing Soviet danger to American preemi-nence in space and to say: "Although we will at this time have to post-pone important parts of our space program, let this fact be clear: We are fixed in our resolve to master the challenge of space."[37]

Johnson and Webb

Yet nothing Johnson said could change the reality of shrinking budgets and flagging enthusiasm for space exploration after Apollo. A *New York Times*'s story in April 1968 observed:

> After a heady decade of uninterrupted hiring, building and dreaming great dreams of far-reaching exploration, the American space program is gearing down to a slower pace and a less certain future. . . . The growing feeling in the space establishment is that once astronauts have landed on the moon, they will have no other place of significance to go for several years because of sharp budget cuts. These cuts have trimmed to the bone all preparations for future missions. It is as if the astronauts are heading for a dead-end on the moon.

By September, after the White House had proposed to reduce NASA spend-ing another quarter of a billion dollars and congressional appropriations committees penciled in only $3.99 billion for NASA in FY 1969, Webb resigned.[38]

There are conflicting accounts of the reasons behind Webb's departure.

Webb himself claimed that he wanted to get out before the *Apollo 7* and *8* missions, the manned orbits of Earth and the Moon in the fall of 1968, so that he could respond to any failure by going after critics in and out of the Congress. Others indicated that Webb was surprised when Johnson accepted his resignation. According to these accounts, Webb had used the threat of resignation repeatedly with the president as a way to press NASA's case. On September 16, 1968, however, Johnson, who had gotten "fed up with this same old story," took Webb up on the offer, saying, "Let's call in the press."[39]

Whatever the realities behind the decision, Webb tried to turn his resignation to NASA's advantage. At a press conference on September 16, he "bitterly" complained that "congressional budget cuts had put the United States second in the space race" behind the Soviets. Though he denied that he was leaving because of reductions totaling $1.4 billion over the previous four years, he nevertheless said that "the agency had been used 'as a sort of whipping boy' by Congress and other agencies competing for Federal funds. And he made it clear that he felt the U.S. is still behind the Soviet Union in space, and that the reason for this second place is a lack of funds." Sources told the *Washington Post* that "Webb was discouraged by the budget cuts, tired of fighting to have Congress restore them and even wearier of debating the urgency of the space program with the Bureau of the Budget and President Johnson."[40]

Webb's public comments provoked an angry response from the administration. Donald F. Hornig, the president's science advisor, sent Johnson a memo describing Webb's assertions as "unconscionable statements" that "were undoubtedly motivated" by NASA's "budgetary problems." Hornig disputed assertions that the Soviets had "'a capability that could change the basic structure and balance of power in the world,' that the U.S. was clearly second in space and that a Soviet manned lunar landing could be achieved in the next year—a time scale that is competitive with, or ahead of Apollo." Hornig thought the United States was at least a year ahead of the Soviets and, if the president agreed, he wished to press Webb and others at NASA to set the record straight. Welsh at the Space Council also felt that Webb's estimate of Soviet space capabilities was "inaccurate" and that the U.S. missions in space had demonstrated American preeminence.[41]

But Johnson sided with Webb, inviting him to respond to Welsh and telling his aides to instruct Hornig not to get into the NASA debate. In a response to Hornig's memo, Johnson said: "Drop it! That is my feeling, but get Jim Webb to get me a prompt reply—all his scientists—all his private ones—to support him and me." In a written memo to Hornig, which

Johnson wanted conveyed only on the phone, the president said, "It is hard for me to believe that Jim Webb would make 'unconscionable statements' or be 'motivated' entirely by budgetary problems." Johnson defended Webb's concern that "the trend of the Soviet program upward and the U.S. program downward" might allow the Soviets "to achieve both the image and reality of power and forward motion." Johnson warned Hornig that "even if your group should develop evidence to sustain their views, your report might be shortly followed by some tragic occurrence in the U.S. program or a major triumph in the Russian one. This would inevitably bring into question the judgment of your group in a way that might impair its usefulness." Johnson also identified himself with Webb's complaints about budget shortfalls, saying, "It was only with great reluctance that for the past two years I have taken action to meet the overall fiscal requirements laid down by a determined group in the Congress by accepting cuts made in the House Appropriations Committee."[42]

Nonetheless, Webb's concerns were greatly exaggerated, as demonstrated by the successful *Apollo 7* and *8* missions in September and December, respectively. But Johnson backed him nevertheless. Partly, he had a warm feeling for Webb, who had served him so loyally for almost five years. To Johnson, this was no small consideration. In February 1968, with the Tet offensive in Vietnam in full swing and the White House besieged by antiwar protests, Johnson talked to Thomas O. Paine about becoming deputy administrator of NASA. As Paine recalled, Johnson stressed the problems faced by his administration, "how much he needed people to come into the government and shoulder part of this burden and relieve him of it, to serve him loyally, help him move the country ahead in these very difficult times." In addition, Johnson liked and admired Webb. After a conversation with Webb about the *Apollo* fire, Johnson remembered telling Mrs. Johnson: "I know now why Jim Webb was an old Marine and a good one. He's got the courage. He goes through a disaster like this and he says, 'We just got to go on and do what we know is right.' . . . And he did."[43]

But more than personal sentiment determined Johnson's support of Webb. He felt that the historical reputation of his administration was in part at stake. If Webb was right about the potential for renewed Soviet dominance in space, if Moscow beat the United States in the Moon race, Johnson believed that he and Webb would be seen as having presided over a failed or at least inadequate space program. By letting Webb beat up on Congress for shortchanging NASA, Johnson was preparing to point the finger at representatives and senators for any retrospective weaknesses historians saw in the Johnson administration's space effort. Shortly after the *Apollo 7* and *8* successes, when Johnson gave Webb NASA's Distinguished Service Medal and praised him as "the best administrator in the Federal Government," he was leaving no doubt that, unlike many in the Congress, he had been an ardent advocate of NASA in general and of Apollo in particular. Johnson wanted to be remembered as a president who made his mark in space.[44]

Conclusion

Johnson's historical reputation as a senator, vice president, and president will never be more than marginally affected by his part in the development of America's space program. His initiatives as an "imperial president" will always identify him more with domestic reforms, such as civil rights, Medicare, federal aid to education, and other war on poverty and Great Society measures, than with Project Apollo. These initiatives will stand with the disaster in Vietnam as the centerpieces of his political career much more than his presidential goals in space. Nevertheless, in time, as the United States progresses into the space age and ever more important discoveries emerge about our universe, Johnson will stand in the front rank of those who had the foresight and determination, as well as the skill to use presidential power—in spite of its undeniable limitations—to initiate America's probe into the farthest reaches of outer space. For this driven, almost madly ambitious man from rural south central Texas, it may be fairly said that in his lifetime he both figuratively and literally reached for the Moon.

Notes

1. This introductory material is from Robert Dallek, *Lone Star Rising: Lyndon Johnson and His Times* (New York: Oxford University Press, 1991), pp. 529–34.

2. See Rowland Evans and Robert Novak, *Lyndon B. Johnson: The Exercise of Power* (New York: Random House, 1966), p. 323; Doris Kearns, *Lyndon Johnson and the American Dream* (New York: Harper and Row, 1976), pp. 162–64; Arthur M. Schlesinger Jr., *Robert Kennedy and His Times* (Boston: Little, Brown, 1978), pp. 622–23.

3. This material is drawn from Lyndon B. Johnson, *The Vantage Point: Perspectives of the Presidency, 1963–1969* (New York: Random House, 1971), pp. 278–79; and Dallek, "The Most Insignificant Office," chap. 1 in *Splendid Misery: Lyndon Johnson and His Times, 1961–1973* (forthcoming from Oxford University Press).

4. Kennedy to Johnson, Apr. 20, 1961; Johnson to Kennedy, Apr. 28, 1961, President's Office Files (POF), John F. Kennedy Library (JFKL), Boston, Mass.; George Reedy, memorandum, n.d. [clearly 1961], beginning: "Despite all the heated public discussion," Office Files, George E. Reedy (OF/GER), Lyndon B. Johnson Library (LBJL), Austin, Texas; Walter A. McDougall, . . . *The Heavens and the Earth: A Political History of the Space Age* (New York: Basic Books, 1985), p. 320; Edward C. Welsh, interview, July 18, 1969, JFKL.

5. McDougall, . . . *The Heavens and the Earth,* pp. 302–3.

6. Ibid., pp. 308–10; Newton Minow, interview, Mar. 19, 1971, LBJL.

7. Kennedy to Johnson, July 29, 1963; Johnson to Kennedy, July 31, 1963, White House Famous Names (WHFN), LBJL; Minow, interview; McDougall, . . . *The Heavens and the Earth,* pp. 322–23, 376, 389–96.

8. Johnson to Welsh, July 26 and 29, 1961; Johnson to Walter Jenkins, Aug. 2, 1961; Welsh to Johnson, Aug. 11, Sept. 8, 13, and 14, 1961; Welsh to Chelsea L. Henson, Oct. 9, 1961, Vice Presidential Papers (VPP), LBJL.

9. George Smathers, interview, Aug. 1–Oct. 24, 1989, Senate Historical Office (SHO), Washington, D.C.; McDougall, . . . *The Heavens and the Earth,* pp. 361–63, 373–76.

10. Richard L. Callaghan, phone conversation with Robert Sherrod, Jan. 28, 1971; Kenneth O'Donnell, phone conversation with Sherrod, May 13, 1971, Robert Sherrod Apollo Collection (RSAC), NASA Historical Reference Collection, NASA History Office, Washington, D.C. Johnson is quoted by Leo Janos in "The Last Days of the President: LBJ in Retirement," *Atlantic,* July 1973, pp. 35–41.

11. Webb, interview by James Burke, May 23, 1979, James Webb Papers, NASA Historical Reference Collection.

12. "Questions from Everett Collier re President Johnson's Attitude on Space," Feb. 8, 1964, White House Central File (WHCF), EX\OS; "Administrative History of NASA" (unpublished manuscript, 1966), pp. 1–9, LBJL; Robert A. Divine, "Lyndon B. Johnson and the Politics of Space," in *The Johnson Years,* vol. 2: *Vietnam, the Environment, and Science,* ed. Robert A. Divine (Lawrence: University Press of Kansas, 1987), pp. 233–34.

13. The Johnson quote is in Divine, *Johnson Years,* p. 234; Kennedy's view is in "Administrative History of NASA," p. 3.

14. Stuart H. Loory, "U.S. Talking Down the Space Race," *New York Herald Tribune,* June 9, 1964, clipping in "NASA Current News folder," NASA Historical Reference Collection; Charles Johnson to McGeorge Bundy, Feb. 4, 1964; Bundy to Johnson, Feb. 29, 1964; Webb to Johnson, Apr. 30 and June 29, 1964; National Security Action Memoranda (NSAM) 285 and 288, National Security Files, LBJL; John Glenn, interview, Oct. 18, 1968, RSAC, NASA Historical Reference Collection.

15. See Merle Miller, *Lyndon* (New York: Simon and Shuster, 1980), p. 410; Kearns, *Lyndon Johnson,* p. 412n3; Vaughn Davis Bornet, *The Presidency of Lyndon B. Johnson* (Lawrence: University Press of Kansas, 1983), p. 8.

16. Webb to Johnson, Dec. 20, 1963, WHCF, GEN, FG260, NASA; Webb to Johnson, Oct. 29, 1965, WHCF, EX, FG260, both in LBJL.

17. See the weekly NASA reports to Johnson from Welsh for 1964 in WHCF, EX\OS, LBJL; and the Webb-Johnson briefing, Oct. 14, 1968, from which the quotations are drawn, in James Webb Papers, NASA Historical Reference Collection.

18. Johnson, "Remarks at the University of Michigan," May 22, 1964, *Public Papers of the Presidents of the United States: Lyndon B. Johnson, 1963–64,* 10 vols. (Washington, D.C.: Government Printing Office, 1965), 1:704–7.

19. Horace Busby to Welsh, May 13, 1964, White House Aides, Horace Busby Papers, LBJL; Webb to Johnson, Nov. 30, 1964; "Introductory Comments," Nov. 12, 1965, quoting Humphrey's speech of Mar. 19, 1965, LBJ Files, NASA Historical Reference Collection.

20. Divine, *Johnson Years,* p. 235; Johnson, interview by Walter Cronkite, July 5, 1969, LBJ Files, NASA Historical Reference Collection; Johnson, *Vantage Point,* pp. 285–86.

21. Welsh to Johnson, May 2, July 21, Sept. 2, and Oct. 24, 1964, WHCF, EX\OS, LBJL; Divine, *Johnson Years,* pp. 234–35.

22. Johnson to Webb, Jan. 30, 1964; Webb to Johnson, Feb. 16, 1965; Jack Valenti to Johnson, Feb. 17, 1965, WHCF, EX\OS; Webb to Johnson, May 20, 1964; Charles S. Sheldon II to George Reedy, May 26, 1964, White House Aides, Bill Moyers Papers, LBJL.

23. Webb to Valenti, Mar. 30, 1965; Valenti to Johnson, Mar. 30 and Apr. 1, 1965; Johnson's note is handwritten on Valenti's March 30 memo, WHCF, EX\OS, LBJL.

24. Johnson, *Vantage Point,* pp. 270–86; Johnson, interview.

25. Welsh, interview; Johnson, interview.

26. Webb to Johnson, Apr. 1 and May 16, 1966, WHCF, Confidential File (CF), OS, LBJL.

27. See "Outer Space Treaty Chronology" in White House Aides, Joseph Califano Papers; and the materials for 1966–67 in WHCF, Legislative Background, Outer Space History, LBJL.

28. Owen to Rostow, Dec. 9, 1966; "Space Goals after Lunar Landing," Oct. 1966, NSF, Subject Files, LBJL.

29. Johnson, interview.

30. See Divine, *Johnson Years,* pp. 238–40; Webb to Johnson, May 16, 1966, WHCF, CF, OS; Webb to Johnson, Aug. 26, 1966, WHCF, EX\OS, LBJL.

31. Schultze to Johnson, Sept. 1 and 20, 1966, WHCF, EX\OS; Webb to Johnson, Dec. 17, 1966, WHCF, EX, FI4\FG260, LBJL; Divine, *Johnson Years,* pp. 240–42.

32. Gallup Poll, *Public Opinion, 1935–1971,* vol. 3, *1959–1971* (New York: Random House, 1972), pp. 1952, 2183–84, 2209; "Poll Finds a Drop in Space Interest," *New York Times,* Dec. 3, 1967, clipping in "NASA Current News folder," Dec. 1967, NASA Historical Reference Collection; *Newsweek* is quoted in "Administrative History of NASA," p. 48.

33. For congressional opinion, see the survey made by Charles Schultze in box 23, WHCF, EX\FI4; Johnson's budget is described in "Administrative History of NASA," pp. 17–19.

34. "Administrative History of NASA," pp. 47–52; Johnson, interview; Senator Clinton P. Anderson, interview by Sherrod, July 25, 1968; Sherrod to John B. Oakes, May 24, 1972, RSAC; Welsh, interview by Eugene M. Emme, Feb. 20, 1969, all in NASA Historical Reference Collection.

35. Webb to Johnson, July 10, 1967, WHCF, CF, Agency Reports, NASA; Webb to Johnson, Aug. 10, 1967, WHCF, EX\FI4\FG200, LBJL; Divine, *Johnson Years,* pp. 243–45.

36. Webb, interview, Apr. 29, 1969, LBJL; Johnson's message to Webb is in Divine, *Johnson Years,* pp. 244–45; the Rayburn quotation is in Dallek, *Lone Star Rising,* p. 427.

37. Webb to Schultze, Nov. 6, 1967, WHCF, EX\FI4, FG260, LBJL.

38. John Noble Wilford, "U.S. Space Effort Is Shrinking after Era of Growth," *New York Times,* Apr. 16, 1968, clipping in "NASA Current News folder," Apr. 1968, NASA Historical Reference Collection; Divine, *Johnson Years,* pp. 238, 245.

39. Webb, interview, Apr. 29, 1969; Thomas O. Paine, interview, Mar. 25, 1969, LBJL; Robert Sherrod, memo on Thomas O. Paine, Aug. 25, 1970; David Williamson Jr., interview by Sherrod, Apr. 10, 1972, James Webb Folder, RSAC, NASA Historical Reference Collection.

40. "Administrative History of NASA," pp. 34–37, 53–55; Carroll Kilpatrick and Thomas O'Toole, "Webb Retires Oct. 7 as U.S. Space Chief," *Washington Post,* Sept. 17, 1968, clipping in "NASA Current News folder," Sept. 1968, NASA Historical Reference Collection.

41. Hornig to Johnson, Sept. 26, 1968; Welsh to Johnson, Sept. 30, 1968, WHCF, EX\OS, LBJL.

42. Webb to Johnson, Oct. 1, 2, and 5, 1968, answering Welsh; Johnson, memorandum, n.d.; Johnson to Hornig, memorandum, n.d., both attached to Larry E. Temple, memorandum for the files, Oct. 10, 1968, WHCF, EX\OS, LBJL.

43. Paine, interview; Johnson, interview.

44. Johnson's awards to and praise for Webb are described in Divine, *Johnson Years,* pp. 246–47.

4

The Presidency, Congress, and the Deceleration of the U.S. Space Program in the 1970s

Joan Hoff

Richard Nixon inherited many things from Lyndon B. Johnson's presidential administration: among them, the Vietnam War, the "Great Society" social reform effort, and the civil space program. In the 1960s, all three experienced spiraling costs, as well as public disagreement of greater or lesser extent over their means and ends, and they all suffered from both managerial problems inside the government and exaggerated expectations by supporters. Each of these difficulties contributed to a growing public dissatisfaction about their purposes and costliness. As the smallest of these inheritances, the space program was the easiest to target for cuts by the new economy-minded administration because it had the least broad public constituency. Three U.S. presidents and three NASA administrations in the 1970s had to face the hard reality that the heyday of NASA's development was over because of previous budgetary and planning problems. Ironically, the dramatic political, cultural, and socioeconomic changes of the tumultuous decade of the 1960s also left this high-tech agency in a time warp not completely of its own making.

By January 1969, when Nixon took office, NASA had already experienced a decline in funding from a peak of $5.25 billion in 1965 to $3.99 billion. The first lunar landings later that year did little to stave off additional funding cuts in future years. In fact, the July and November Moon landings probably contributed to, rather than diminished, the "three-ring circus" atmosphere, and hence the disillusionment felt by so many Americans about Project Apollo. The program, while technologically innovative and visually exciting, left much to be desired from most other vantage points. Many Americans felt little sympathy for the celebrities who became regular members of an elite audience at Apollo launches, for astronauts promoting all types of business endeavors and marketing space memorabilia, for NASA leaders who by 1968–69 seemed to believe (in the wake of the successful lunar missions) that their agency deserved whatever funding it requested, for rising taxes and a worsening economic situation that were exacerbated by these programs, or finally for a worried aerospace industry backed by major universities that stood to lose billions of dollars if the space program were cut. By 1969, there were simply too many charges of "misplaced government priorities" and "misguided government allocations of funds" for NASA to continue with business as usual.[1]

For all the blame laid on the Nixon administration because of Watergate, one positive thing he did was to move space technology away from being merely a political and military weapon in the cold war—as it had usually been conceived since the successful Soviet launching of *Sputnik I* on October 4, 1957—toward a more balanced and deliberate effort that avoided international competition. Nixon tried to downplay the cold war tensions that had done so much to make Project Apollo the vehicle for achieving international prestige and to return to the more ordered approach of his 1950s mentor, Dwight D. Eisenhower.[2] Nixon was concerned about scientific-technological programs that might stress engineering over science, competition over cooperation, civilian over military, and adventure over applications, and this consideration prompted him to deal carefully with NASA's programs.[3] Equally important, Nixon's emphasis on frugality in government spending prompted caution on his part in endorsing any effort in space. The result was a deceleration of the rate of space exploration in the 1970s, an emphasis on scientific return, and a commitment to obtaining the most efficient space effort for the least expenditure of funds.

Even had Nixon been so inclined, he probably could not have continued his predecessor's impossible dream of capturing outer space from the Soviets as a twentieth-century equivalent to the road system of the Roman Empire or control of the seas by the British Navy.[4] In fact, in an exuberant post-presidency interview with Walter Cronkite after the first Moon landing, Johnson attributed the flood tide of social legislation that became the hallmark of his administration to the fact that the space program had begun it all by breaking down the resistance in the South among Democrats and Republicans alike to federally funded programs.[5]

Nixon, while not above wining and dining astronauts as American heroes to further his political purposes, never exhibited the personal enthusiasm for or expansive commitment to the space program that Johnson and John F. Kennedy had shown. This was probably in part because he did not need to use the space program to prove himself able to deal with the Soviets, as Kennedy and Johnson apparently thought they did. Moreover, Nixon had inherited too many economic problems created by the massive spending programs launched in the name of the cold war in Vietnam and the Great Society. These programs, in response to crisis and division and a breakdown of the older political tradition, had gradually lost support among the public in the course of the 1960s.[6] In a word, fighting the cold war and conducting a domestic reform program had become so expensive that the Nixon administration had no choice but to retrench.

There were other concrete reasons for the continued deceleration than just national polls showing opposition to the expense of the manned space program,[7] or what NASA Administrator James C. Fletcher called the "antitechnology kick" of the countercultural generation,[8] or even the antidemocratic overtones and cultural elitism of the technocratic approach to government introduced by the Kennedy administration.[9] While these attitudes made future funding of the space program more difficult than in the years between *Sputnik I* and the mid-1960s, I believe that they are incomplete answers and that to them must be added an institutional obstinacy at NASA when asked to comply with changing government budgeting methods and changing public expectations about the meaning of the space program by the late 1960s and throughout the 1970s.

The principal reasons for the deceleration of the space program under Nixon and for the rest of the 1970s arose from four rather broad issues that have been largely unexplored in the history of NASA: personnel, budgetary, foreign policy, and political factors. By personnel, I mean that Nixon had no close advisors promoting the space program as he did on the major domestic initiatives he undertook. Put most simply, NASA Administrators Thomas O. Paine and James C. Fletcher, and even Nixon's first two science and technology advisors, Lee A. DuBridge and Edward E. David Jr., did not have the ear of the president or any of Nixon's inner staff.[10] At the time, Nixon also did not perceive the space program in crisis due to lower funding. On other domestic issues, this "crisis mentality" on the president's part had proven an essential criterion if additional funding was to be recommended. In particular, I am referring here to the environmental and welfare legislation proposed during Nixon's first administration. While Nixon thought that a crisis existed in funding research and development, he did not think a similar one existed in the space program.[11]

From a strictly budgetary standpoint, NASA was a classic example of the myriad cost overruns present throughout the federal government in the first half of the 1960s.[12] As an example, in 1963 James E. Webb announced the establishment of the Electronics Research Center in Boston, one of his most criticized administrative decisions. The subsequent investigation of this and a number of other governmental procurement decisions by Webb continued into the last half of the 1960s, yet one would never know this from the memoranda and subsequent budgets submitted by Paine, who succeeded Webb in September 1968.[13] Neither Paine nor his successor, Fletcher, seemed to grasp the necessity of not only complying with, but actually understanding the new cost accounting methods instituted by the Johnson, Nixon, and Carter administrations. Neither grasped the importance of knowing with whom in the Bureau of the Budget (BOB) and later the Office of Management and Budget (OMB) they absolutely had to maintain relations in order to receive serious consideration for their projects during the complicated process that went into determining the yearly expenses of government. As I will detail later in this chapter, Paine's behavior during the budget process of 1970–71, in particular, appeared to Nixon stalwarts at best as irrational and at worst as obstinately arrogant.

My third point is that the space budget would probably have been reduced in any case in the 1970s because it had originated as a product of the cold war and was therefore subject to rising and falling expectations about favorable U.S.-Soviet relations. Under Nixon these expectations were high, and therefore arguments about "beating" the Soviets in space carried less weight that they had under Kennedy and Johnson, but NASA administrators and White House science advisors in 1969–72 failed to appreciate this important shift, and so they tried to convince Nixon to commit himself to certain aspects of the space program before the Soviets did.[14] In fact, instead of funding more competition with the Soviets, Nixon's geopolitical ideas and his policy of détente emphasized international cooperation and cooptation of the Soviet Union. This included matters involving space.

Finally, the political considerations that worked against increased funding for NASA are self-evident. By 1969, liberals and conservatives in both parties, but especially liberal Democrats, were highly critical of more spending for space when such domestic problems as the environment, poverty, crime, urban renewal, and racism loomed large. Given the fact that until 1988 Nixon remained the only twentieth-century president to be elected without his party having control of either house of Congress, he was constantly trying to co-opt liberal opinion on certain issues, such as welfare, to minimize liberal opposition to the war. Likewise, he initially tried to placate conservatives with a "southern strategy," as demonstrated through his first unsuccessful Supreme Court nominees, only to find that they did not fall in line with him on social policy issues.[15] When the chair of the House Committee on Science and Astronautics, George P. Miller (D-CA) called the manned Mars mission "premature," and the chair of the Senate Aeronautical and Space Sciences Committee, Clinton P. Anderson (D-NM), did not think "we could afford it now," echoing the skepticism of other "key congressional leaders, Nixon and his close advisors became convinced they would lose little in Congress or the country at large by trimming the NASA budget."[16] The same reasoning prevailed for the rest of the decade under Ford and Carter.

I will now take up each of these four perspectives in more detail to argue that even if the country had been able to finance the cold war in the style to which it had become accustomed under Kennedy and Johnson, and even without the economic dislocation such financing was causing the American economy by 1969, Nixon in all likelihood would not have continued to fund NASA at its peak of the mid-1960s. Indeed, that Nixon would have increased NASA's budget to its 1965 level is difficult to imagine, for reasons having to do with his close advisors and their relationship with key NASA personnel, reorganization of the executive branch as it affected the budget process, Nixon's "grand design" for foreign policy which included détente with the Soviet Union, and the president's relations with Congress over other domestic and foreign policy issues, in addition to political disagreement among members of Congress over the space program in the post-Apollo era. Moreover, Nixon felt the need to cut NASA's budget even further. Many of these same conditions (with the exception of détente with the Soviet Union) prevailed under Gerald R. Ford's and Jimmy Carter's administrations as well, ensuring that NASA funding throughout the 1970s would not return to the heights it reached trying to beat the Soviets to the Moon.

Nixon's Advisors and NASA: A Gap That Was Never Closed

Before describing the specific attitudes about NASA among Nixon's advisors, I want to consider the president's own views about space. In general, he was probably not a space buff or very knowledgeable about technical details of Project Apollo, which he had inherited at the moment of its dramatic Moon walks. Moreover, Nixon's *Memoirs* reveal no particular interest in the subject as a member of Congress or later as vice president.[17] It is true that under Eisenhower he had been so impressed with *Sputnik* that he countered the statements of such presidential aides as Sherman Adams who said the satellite race was no more than an "an outer space basketball game," and Nixon advocated increased spending for the missile program and later for human spaceflight vehicles. In general, however, he seems to have subscribed to the more cautious way that Eisenhower approached the militarization of space by connecting it with defense, rather than civilian engineering and prestige.[18]

In fact, Nixon mentioned the space program more during his 1960 campaign for the presidency than he did in the 1968 campaign because it did not become a cold war debating issue between him and Hubert Humphrey as it had between him and Kennedy. In 1968, he stressed increasing federal and private funds for research and development (R&D) for civilian needs more than space research, and as president he fulfilled that goal. By 1972, the Nixon administration had increased R&D funding from $15.6 billion in FY 1969 to $17.8 billion for FY 1973, or an increase of 14 percent.[19] Although Nixon spoke of the *Apollo 11* mission as the "most exiting event of the first year of my presidency," his presidential papers document clearly that his personal interest was more in the diplomacy of space and in the defense and commercial applications of spinoffs of the space program, such as the antiballistic missile (ABM) system and supersonic transportation, than in its purely scientific or interplanetary potential.[20] In one of his first meetings with Paine, Nixon kept repeating the phrase "space and defense," leaving little doubt in the administrator's mind that they were inextricably connected in the president's.[21]

As with all issues, Nixon did his homework and tried to keep informed, but NASA would probably not even have been on his list of priorities for study had it not been that Johnson intentionally left it for him to formulate national space policy in the post-Apollo period. Again, for someone who has studied other aspects of the Nixon administration, this is unsurprising, because Johnson also deliberately postponed implementing desegregation of southern schools so that this controversial task would likewise fall on the Nixon watch.[22] Had it not been for Johnson's procrastination, Nixon would not have immediately turned his attention to space policy by establishing during the interregnum period a task force headed by Charles Townes. Its January report to the president-elect essentially reiterated what the 1967 President's Science Advisory Committee (PSAC) had told Johnson. In both reports, the "code word" became "balance," meaning a "program based on the expectation of eventual manned planetary exploration, integrating manned and unmanned efforts," with the specific recommendation from the Townes task force— which NASA opposed—that a $4 billion budget would be "adequate for the important programs envisaged." Of course, had Nixon personally wanted a more aggressive space program he could have forced the issue and countermanded the Townes report. That he did not do so indicates his basic agreement with a moderate level of spending in this arena and his preoccupation with other issues. As a result, like the arbitrary figure that one of Nixon's other task forces set for a new welfare program, the $4 billion NASA budget became embedded in the thinking of Nixon's White House advisors.[23]

This task force report led Nixon to ask DuBridge, his first presidential science advisor and director of the Office of Science and Technology (OST), to establish a Space Task Group (STG) headed by Vice President Spiro T. Agnew (as chair of the Space Council) to "report on possible cost reductions in specified portions of our space program." In another memo, Nixon insisted that he wanted a "definitive recommendation on the direction which the U.S. space program should take in the post-Apollo period," specifically, "a coordinated program and budget proposal," as well as information on "international implications and cooperation."[24] This directive proved a mixed blessing for NASA because Paine assumed almost immediately that Agnew's personal and public support of a "manned flight to Mars by the end of this century" would carry the day inside the White House and BOB, when nothing could have been further from the truth. Agnew carried little weight with Nixon or his close advisors and none with the BOB director, Robert Mayo, whom the vice president purportedly called a "cheapskate" at one of the STG meetings.[25] Consequently, Paine wasted much time and effort in the summer and fall of 1969 acting as a link between NASA and the STG in the hope of using this report as ammunition against Mayo, instead of preparing the reports requested by the BOB for FY 1971. He apparently never understood the limited function and impact of most task force reports, and he certainly misjudged Agnew's influence on the Nixon administration.[26]

Even in the best of times, but particularly in the time of turmoil that existed in the late 1960s and early 1970s, presidential policies seldom reflect exclusively the ideas or personality of any given president. They are, instead, the much more collective product of his aides and various divisions of the executive branch and his own personal administrative mode of operation. I have divided those who advised Nixon on major issues into two camps: the "free-thinking" outsiders who brainstormed with the president about new ideas and comprehensive programs and the "political-broker" insiders who worked to draft and implement his legislative and administrative priorities.[27] Neither of these two sets of advisors included any outspoken advocates of the space program and, therefore, neither took it upon itself to present the space program or the NASA budget as a high-priority, crisis issue to the president.

These two quite different sets of advisors agreed on one thing: the president should appoint generalists (policy specialists and politicians) like themselves to oversee the work of civil service specialists (experts or technicians) from the very beginning of his presidency. Their reasoning was that generalists would provide him with more moderate and less self- (or agency-) interested advice. Initially, however, Nixon thought that he could appoint generalists both as strong agency and department heads and as strong White House staff people to monitor the specialists. After the insiders and outsiders inevitably clashed in the course of his first administration, he decided to move members of his White House staff (and other generalists who had proven loyalists on policy) from his personal staff into key positions within the executive branch.[28] This did not mean that Nixon was against technology or brought an antitechnology bias to the White House. He and his two sets of advisors simply did not want technocrats to be in the influential policy-making positions they had occupied under Kennedy and Johnson.

In a White House atmosphere in which there was not strong presidential interest in either NASA or the civil space program, few on the president's staff showed much real interest either. John D. Ehrlichman's papers reveal that, although he emerged as the strongest (and one of the most liberal) of Nixon's insider advisors on domestic policy, he had little interest in the space program. (In one interview, Ehrlichman implied that Nixon's major interest in Apollo was as a vehicle for uplifting pomp and circumstance for the nation similar to the panoplies surrounding the return of the POWs, national parades, and the short-lived experiment with palace guard uniforms for the White House police force.)[29] This meant that Peter M. Flanigan, an investment banker who had been the deputy campaign manager for Nixon in 1968, was assigned oversight responsibilities for space as part of his general duties as assistant to the president for internal economic affairs. Flanigan in turn relied on Thomas Clay Whitehead, a former RAND systems analyst, to evaluate NASA budget and planning proposals. Although Jerome Wolff, an aide Agnew brought from Maryland to advise him on science and technology, contacted all these White House advisors about the STG report, Flanigan and Whitehead turned out to be the critical influences in making decisions about NASA's FY 1971 budget.[30]

In this environment, there was little push from Nixon insiders for an aggressive space program. Increased funding for NASA would have been an uphill battle in any case, however, since neither DuBridge nor David as sci-

ence advisors favored human spaceflight programs and Flanigan and Whitehead were mainly interested in proving to the president that they were at
least as committed to cost-effectiveness and to producing a balanced budget as was Mayo at the BOB. In truth, no one in the White House had much
interest in the space program or wanted to increase its funding levels. As a
result, in February 1969 Paine tried to convince Nixon that he should issue
a "general directive to define the future goals of [the] manned space flight"
program "prior to the final decisions on plans that will be recommended to
you . . . by . . . the Task Force." Paine did so on the grounds that the Soviets were about to announce their own plans in the "orbital space station
field" during the summer months and "that would take the edge off [Nix-
on's] announcement of a similar U.S. objective in the fall."[31]

When Nixon's advisors told him not to respond to Paine's attempted
end run, the administrator threw all of his energy into influencing the STG
report and in the process systematically offended both White House aides
and top BOB officials. Whether he felt compelled to act in this manner because, as a Democrat and member of the NAACP, he had actively supported the campaign of Hubert Humphrey, or whether this was his normal operating style remains debatable. His resignation in September 1970 came
as a welcome relief to both the executive and legislative branches of government. One of the reasons for the five-month delay in finding his successor was that his behavior had convinced the Nixon administration that
it did not want another Paine as head of NASA. Flanigan, for example,
was specifically told to find a NASA administrator "who will turn down
NASA's empire-building fervor and turn his attention to 1) sensible straight-
ening away of internal management and 2) working *with* OMB and White
House."[32]

While tempers improved once Fletcher became NASA administrator in
the spring of 1971, the agency's funding situation did not. This was in part
because Fletcher relied too heavily on Flanigan for access to Nixon, and
by the time of Fletcher's arrival Flanigan had also been appointed executive director of the Council on International Economic Policy (CIEP) and
was too busy to be a space advocate inside the White House even had he
been so inclined. Fletcher's right-hand man on space, Thomas Whitehead,
also became involved in other policy matters, informing the administrator that Paine could not rely on him as in the past. Because of the situation in the Nixon White House in the early 1970s, without Ehrlichman's
active support, cuts in the NASA budget could not have been prevented,
let alone the increases, which Paine demanded and Fletcher pleaded for,
achieved. In a January 1970 meeting with Paine, Nixon told him that he

regretted the additional cuts in FY 1971 but that Congress and the people were in favor of severe cuts in "space and defense."[33] By that time, Nixon had already begun to withdraw U.S. troops from Vietnam and cut back on defense spending. Whether the public and Congress would have tolerated some increase in spending for space for purely scientific purposes unrelated to defense was never tested by Nixon. For the president, defense and space were one and same in the budget.

From BOB to OMB and More Headaches for NASA

Closely related to this personnel problem, which NASA never solved under Nixon and only marginally so under Presidents Ford and Carter, with whom Fletcher had more cordial and direct access, was the budgetary process itself. In the late 1960s and early 1970s, both the Johnson and Nixon administrations introduced new concepts into budget formulation. Nixon's ideas were particularly striking because with congressional approval they transformed the Bureau of the Budget into the Office of Management and Budget in July 1970. This reorganization was based in part on the advanced corporate theory known as "management by objectives" (MBO) recommended by the President's Advisory Council on Executive Organization (PACEO), also known as the Ash Council. Nixon did not introduce the MBO component until early 1973, the same year he eliminated the Office of Science and Technology (OST), saying that the National Science Foundation (NSF) was better equipped to carry out the advisory functions of the White House science advisor. Both actions were based on recommendations the Ash Council had made as early as the fall of 1969, which were in turn based on corporate organization theories.[34]

With or without MBO, the establishment of the OMB remains one of the most influential management changes initiated by Nixon. The OMB's review procedures, based on central clearance of all department, agency, and commission budgets *before* they were submitted to Congress, was institutionalized by Nixon's successors because of its power to evaluate program performance and to control spending.[35] The OMB can be viewed as his most "imperialistic" achievement because "OMB is on paper the single most powerful managerial unit in the government." It has been so significant that since 1973 only the CIA and the Defense Department have successfully challenged OMB's budget-setting powers. In spite of this, by the early 1970s only a few isolated scientists had realized the OMB's potential for "usurping scientific judgment and congressional intent through its impoundments of allocations for scientific research."[36]

Neither move reflected any real antitechnological or antiscience bias on the part of the president or his close advisors, including Roy L. Ash who at the time was president of the high-tech Litton Industries. Although congressional hearings in 1973–74 on Nixon's Reorganization Plan No. 1 were influenced by the emerging Watergate scandal and cannot be taken at face value in discussing the wisdom of eliminating the OST and PSAC, the idea of using the National Science Foundation because it fostered "pluralism" in government funding for science (and hence, better geographical distribution of funds) had its defenders among some scientists, as well some business theorists. (Later, as head of the OMB, Ash favored funding for certain projects of the NSF over those of NASA.) This reorganization was also completely in keeping with other decentralized programs established in the name of Nixon's New Federalism, but some scientists viewed the demise of OST and PSAC as depriving "the science community of substantial status and influence in the White House," not realizing how little influence either had under the presidents since Kennedy as personal White House advisors came to play increasingly important roles.[37]

Likewise, the adoption of management by objective reflected no intrinsic bias against science by the Nixon administration. This recommendation from the Ash Council came on the heels of the failure of the Performance Measures System launched in 1971 as a variation of the Planning Programming Budgeting System (PPBS), originally introduced in 1965 under the Johnson administration. The MBO system was not intended to save money, to decide between competing programs, or even to be a means for the White House to mandate priorities for individual departments. However, because it insisted on maximizing the best use of government funding based on national priorities, MBO indirectly allowed any president more influence in facilitating the achievement of some of his preconceived objectives.[38]

There is little evidence in Nixon's presidential papers or in NASA records, however, that the agency understood the enormous significance of the basic structural and analytical change that had taken place in budget formulation by the summer of 1970 or the role of OMB in the process. This is especially evident in all the interviews of Fletcher, as well as his correspondence, in which he expressed either dismay or irritation with the OMB's procedures but little understanding of how to "play" the game of compliance so as not to hurt NASA requests for budget increases. The same appears true of his attitude toward Congress.[39]

This was even more true of NASA's reaction to Carter's famous—or infamous depending on your point of view—"zero-based budget" (ZBB) plans for the OMB. Fletcher wrote a rather pathetic note, saying: "I am not sure what 'zero-based budget' means—but what it used to mean is what I thought we were doing every year. Is this going to give us problems?"[40] It is also not evident that Robert A. Frosch—the oceanographer, flutist, and sculptor who became Carter's NASA administrator in 1977—was prepared to present NASA budgets any more effectively under OMB management than had his predecessors. Not since Webb in the 1960s had NASA effectively made its case for large budget growth—Webb even received praise from Congress for "his ability to present a very complex budget every year with the enthusiasm of a true believer"—but Webb had operated in a much different budget environment, which he understood from his time as Truman's head of the BOB.[41]

Of the three NASA administrators during the 1970s, Paine behaved more arrogantly than the others, especially when it came to conforming to the budgetary process. He ignored the BOB's requests for PPBS budget analysis not once but twice in the spring and fall of 1969. That Paine's efforts were poor in this regard was evident from the moment that he tried to comply with the requests from the STG for its long-range plans. Webb did not put any formal long-range planning structure in place until 1968 even though the House Committee on Science and Technology had called for a report from NASA by the end of 1966 on future national space objectives. According to one government study, it is quite possible that, if Webb had taken this congressional request seriously, NASA might have fared better later in the budget-making process. Instead, Webb dismissed this request, countering, "Because of the difficult budgetary situation resulting from the war in Vietnam and other factors . . . we [are] precluded by the regular budgetary procedures from presenting specific statements on our future plans at this time." Thus, NASA began to use spending for Vietnam and domestic social programs to rationalize its own inability to present coherent plans or budgets for the future of spaceflight. In 1979, members of the House Committee on Science and Technology were still complaining about a "lack of long-range planning and what seems to be a lack of more specificity in what may be the plans [of NASA] for the future."[42]

Not until 1968 did Webb belatedly put Homer E. Newell, who had run the NASA space science program, in charge of a formal planning structure. Apparently, Newell operated under the illusion that post-Apollo policy would arise out of some kind of public debate and that NASA would simply follow that lead. When this curious way of approaching long-range planning did not materialize and the BOB requested that NASA establish a PPBS system on which to base future budgets, Newell proceeded to set up a cumbersome and unworkable structure, consisting of a Planning Steering Group (PSG) and twelve planning panels representing the complicated competition among manned and unmanned subunits of NASA.[43]

Not surprisingly, NASA's PSG produced mountains of data, but no coherent plan emerged from the process that satisfied Paine in 1969 so that he could respond to the STG request. So Paine turned for help to George E. Mueller, head of the Office of Manned Space Flight. Mueller produced what was called the "integrated plan," calling for cost-effectiveness through developing a reusable spacecraft for operations between the Moon and Earth, once again linking NASA's future to a huge human spaceflight project, as had been the case during Project Apollo. Paine liked Mueller's general idea but arbitrarily decided to throw out his "cislunar" emphasis and replace it with human planetary exploration, which would be more inclusive and therefore appeal to more scientists through the pursuit of a larger goal of solar system exploration. To help focus this plan, Paine highlighted a manned Mars expedition in the report that went to the STG a day before the *Apollo 11* launch on July 16, 1969. In spite of the hoopla surrounding the lunar landing, within days Congress and the public were questioning the cost of placing an astronaut on Mars as advocated by the STG.[44]

In the long run, Paine's arbitrary decision to stress a mission to Mars was unsound, especially since Nixon had privately and publicly stressed international cooperation in space based on a "partner instead of a patron" relationship. The president's greatest worry was that congressional opponents of the space program would negatively compare "his positive statements on space to problems in poverty and social programs here on earth."[45] While Paine sympathized with this view, he hindered rather than helped the Nixon administration in 1969–70 with his bullish attitude on such matters as a U.S. mission to Mars in the near term.[46]

In the process of working on its report, the STG not only contacted members of Congress but also prominent individual Americans called "invited Contributors."[47] Among them was Shirley Temple Black, who sent the vice president a thoughtful nine-page report, stressing international cooperation as the highest priority in space. The vast majority of the members of the STG, the PSAC, and the outside contributors opposed Paine's idea of a hastily organized Mars expedition based on current technological capabilities. Most importantly, the STG member Robert C. Seamans Jr., secretary of the Air Force, joined by Mayo, the budget director, strongly opposed a manned planetary mission.[48] By the end of July, both NASA and the PSAC presented reports to the STG. Criticism in Congress and by the

American public led the STG to recommend the concurrent or sequential building of a space station and shuttle and to speak only of an "eventual, potential option of [a] manned mission to Mars before the year 2000." In keeping with White House advisors' recommendations, the report stressed the low- rather than high-cost options that would not cause the president political damage if he rejected any of them. Therefore, the STG did not recommend any one of the three specific program options to the president; neither were there figures in the report analyzing the various costs of the alternatives.[49] In essence, the STG report represented much ado about nothing, except that Paine mistakenly chose to make it the focus of his activities throughout much of 1969.

Given the inefficient budget process and Paine's obsession with satisfying requests from the Space Task Group rather than those from the BOB, he could not comply with two BOB deadlines for NASA's FY 1971 budget submissions in the spring and fall of 1969. It also did not help matters when Mayo criticized the "shortcomings" of the STG report, since it did not recommend any one program to the president. To many observers in the White House and BOB, the NASA budget process was in a state of "disarray," indicative of the agency's inability to put together the type of budget being requested by the budget director and his staff.[50]

After declaring that the inevitable BOB cuts were "unacceptable," Paine appealed the decision and then in November 1969 presented a budget of $4.25 billion (down from an earlier $4.497 billion request) that he said was the lowest the space agency could tolerate. The BOB responded that NASA's budget could be no more than $3.7 billion (up from an original mark of $3.349 billion). Obviously, an impasse had been reached; at this point, the White House staff intervened, but not on the side of NASA because only Vice President Agnew, whose advice was ignored, supported a higher budget. As a result, NASA had to accept not only a cut in FY 1970 prepared by the Johnson administration after the Nixon administration reviewed it but also in the BOB figure for FY 1971 after the White House staff, consisting of Flanigan and Whitehead, had reviewed that one and recommended $3.53 billion. Even as Paine was announcing this figure to the press, the White House decided on another 2.5 percent across-the-board cut for all agencies in order to present a balanced budget to Congress. Thus, without consultation with NASA, the agency's budget was reduced to $3.3 billion.[51]

What these and subsequent figures throughout the Nixon and Ford administrations meant was (1) no development of a space station or space shuttle during FY 1971, (2) a reduction of Apollo missions from three to one per year, (3) the termination of the Saturn V booster, and (4) no new unmanned projects because science and application programs would be held to existing levels.[52] In this fashion, the budget begat space policy instead of space policy begetting budget, as had been the case during the heyday of Apollo in the Kennedy and Johnson administrations.

Early in the 1976 presidential campaign, Carter tried to distance himself from Nixon's and Ford's "balanced" approach to manned and unmanned space projects, saying his administration would reject "costly missions . . . in favor of unmanned scientific exploration and practical applications of existing technology."[53] In particular, Carter thought that it was "neither feasible nor necessary at this time to commit the U.S. to a high-challenge space engineering initiative comparable to Apollo." Unlike Kennedy and Nixon, Carter did not appoint an interregnum task force to study U.S. space policy. Instead, he relied on a summary by one individual—Nick MacNeil. The National Security Council conducted the only comprehensive review of space policy under Carter, indicating how closely his administration linked space and defense. As president, Carter returned U.S. space policy closer to what it had been under Eisenhower, who "saw the development of space technology only as a means, not as an end in itself." Specifically, Carter said (somewhat redundantly) that "activities will be pursued in space when it appears that national objectives can most efficiently be met through space activities." He also decided to retain the same link between "military and space" that Nixon brought to the office of the presidency. And, of course, so did Reagan with his Strategic Defense Initiative program.[54]

This acrimonious relationship between NASA, the White House, and the BOB preceded both Fletcher's appointment as NASA administrator in 1971 and Nixon's decision to endorse the reusable space shuttle program in 1972. By that time, however, Mayo and the BOB were no longer around to plague NASA; instead there was the new Office of Management and Budget under the direction of Ash, whose earlier reorganization plans as head of PACEO had contained implicit criticisms of NASA's management approach. Paine had left office after adopting the most excessive aspects of Vice President Agnew's argot by taking potshots at "Potland" (a reference to those in the counterculture whom Paine insisted were enemies of technology). This did not endear him to Nixon stalwarts. Neither did his antics in a 1970 commencement address at Worcester Polytechnic Institute when Paine proposed a hypothetical cabinet made of up of Timothy Leary as secretary of agriculture, Jane Fonda as secretary of interior, Arlo Guthrie as secretary of health, education, and welfare, Ralph Nader as secretary of commerce, and Bobby Seale as attorney general. Paine also took aim at congressional critics of the space program, such as Senator Edward M. Kennedy (D-MA), accusing him of poor taste in cuisine because of the quality of a lunch he had with him. Without question, Paine bequeathed to Fletcher a bewildering public relations problem with his trail of mixed and ill-considered messages, leading one commentator to quip that Paine "appear[ed] a little fey." Paine also left office under the cloud of mishandling a $50 million contract with GE-Hiller Fairchild. With all of these concerns, there was little sentiment in the Nixon administration for Paine to stay at NASA.[55]

After all that had gone on before, Fletcher was ridiculed in 1971 when he took over NASA's reins—one reporter even misleadingly described him as a "Mormon for the Moon"—and he had to work hard to return to a more productive relationship with the White House by adopting a mild-mannered demeanor in dealings with the Nixon, Ford, and Carter administrations. The press inaccurately assumed that Fletcher would not stay long in the job because he only took a two-year leave of absence as president of the University of Utah. He stayed at NASA from 1971 until 1977—almost as long as Webb—and he became quite successful in working quietly to achieve the ends of the agency as he interpreted them. Nonetheless, like his predecessors, Fletcher felt defensive toward the OMB and its budget processes. He once noted that the one thing he had learned as NASA administrator was that a "deal from OMB is no deal at all," in reference to a perception that NASA had suffered budget cuts greater than the level agreed to by the agency and the White House. He also criticized Congress for having too few "pro-space" members on the appropriations committees.[56]

Fletcher had to deal immediately with Nixon's March 7, 1970, statement on space that specifically encouraged "greater international cooperation" in keeping with his September 1969 address to the United Nations in which the president called for the "internationalization of man's epic venture into space."[57] Because this was such a vague mandate, Nixon's White House advisors, the OMB, and NASA all tried to convince the president that their specific recommendations should be selected to fill in the gaps. In this process, two of Nixon's political-broker insider advisors at OMB—Caspar W. Weinberger, OMB deputy director, and Donald B. Rice, an OMB assistant in charge of NASA budget initiatives—provided Nixon with most of his information before his 1972 decision in favor of the reusable space shuttle. Within the White House, Flanigan and Ehrlichman also occasionally contributed to the discussion.[58]

Rice carried on Mayo's budget slashing under his new boss, Ash, at the newly created Office of Management and Budget. In particular, Rice considered NASA incapable of qualitatively evaluating its own programs and priorities. No other federal agency on the domestic side of policy making except the Atomic Energy Commission, according to Rice, was so unreflective. NASA always appeared to be trying to place the president in an either/or situation. As Rice put it, the president "could either proceed with [NASA's] program . . . or he could take the U.S. out of manned space." Rice kept pressing NASA to produce the best shuttle for the least money. Fletcher tried to convince him of NASA's ability to conduct a space program that included a reusable space shuttle. He eventually did so, compromising NASA's plans for full reusability in the process, but not before much more ill will had been generated between NASA and the OMB in 1970–71.[59]

Had it not been for Weinberger's presence, and to a lesser degree that of Robert C. Seamans and David Packard at the Department of Defense and a few individuals involved in foreign policy issues such as Henry Kissinger—all of whom also supported the reusable shuttle idea, but for quite different reasons—the space shuttle decision might not have been reached in 1972. It is to Fletcher's credit that he realized this and incorporated and kept key leaders in the Nixon administration in his shuttle coalition. Among other selling points, he emphasized military applications and the international cooperation inherent in the space shuttle program; several of those leaders—including Nixon—were impressed by both arguments.[60]

Weinberger opposed recommendations from the OMB staffers who did not want to fund the shuttle program. In a crucial memorandum to Nixon in August 1971, the OMB deputy director argued that the administration should not give the world the impression that America's best years in space were past; so he recommended funding the last two Apollo flights, the *Skylab* orbital workshop, and the space shuttle. Weinberger's access to Nixon through his second boss at OMB, George Shultz, may have "saved" NASA from declining even further as a budgetary priority, because Nixon replied in a handwritten comment: "I agree with Cap."[61] OMB staffers and NASA personnel were not immediately informed of this exchange, and they continued to haggle over funding for space, including the space shuttle. At one point, the White House science advisor, Ed David, informed Fletcher that there were no staffers "in OMB who could be completely trusted—not that they were dishonest, but that their sole function was to put a ratchet on the budget and [they] couldn't make a commitment to hold the line on anything."[62]

An important factor aiding NASA in gaining approval of the space shuttle at this time was the fact that 1972 was, after all, an election year. Ehr-lichman pointed out to Nixon that some "close" states controlling large number of electoral votes were also those with space industries that would benefit from the new space shuttle program. Toward the end of the pro-cess, in December 1971, Fletcher and George M. Low, NASA deputy ad-ministrator, met with Flanigan and David. At that time, the NASA people were told that the president had all but decided to go ahead with the shuttle program. It was then that Fletcher and Low flew to California to meet with Nixon when he announced his decision on January 5, 1972.[63] The deci-sion-making process had clearly been driven by advisors who knew Nix-on best, not NASA personnel.

In retrospect, it is clear that Nixon had no choice but to opt for *some* kind of major human spaceflight project to succeed Apollo; the astronauts provided the necessary human element of a science that was largely unin-telligible to the average person. No president in the 1970s wanted to be responsible for "killing" the only compassionate symbol of space explo-ration: the astronaut in orbit. Nixon, Ford, and Carter proved no excep-tions to this rule, so the U.S. manned space program continued but at a decelerated pace, except in the area of space diplomacy.

The Diplomacy of Space under Nixon: Cooperation Not Competition

During the summer of 1969 as U.S. foreign policy was being formulated largely in secret (including the bombing of Cambodia), congressional op-position preoccupied the president. The ways in which he and his aides tried to outmaneuver diplomatic initiatives on the part of the U.S. Senate forced Nixon into a delicate political balancing act that ultimately shaped his and Henry Kissinger's "grand design" more than they wanted at the time and more than they have admitted since. Rather than a blueprint for U.S. diplomacy and much like the détente that became its centerpiece, it remained a process rather than a fixed policy.[64]

In reality, détente from a "Nixinger" perspective represented nothing more or less than the political and economic means, strategy, or process (as opposed to an actual goal or condition) for (1) avoiding nuclear war, (2) "building a network of mutually advantageous relationships," and (3) modifying Soviet behavior by gaining its de facto acceptance of interna-tional cooperation and competition (sometimes referred to as "competi-tive coexistence") in order to preserve international stability by accord-ing the Soviet Union a greater stake in the status quo. In other words, it was an attempt to co-opt the USSR. To a lesser degree than some have argued, détente also reflected the domestic and international economic problems the United States faced as a result of the impact of the Vietnam War, which meant among other things that the country could not continue to fund the race for space with the Soviets at previous levels.[65] One aspect of détente not explained to the American people was that for Nixon space always meant defense first; he associated it with ICBMs, reconnaissance satellites, and, an especially important personal agenda item, the antiballistic missile (ABM) program.

Soon after Nixon's inauguration, the U.S. space program entered this equation in a very unusual way. He viewed any congressional opposition to the proposed ABM system as not only a threat to the possibility of détente but also to continued U.S. conventional arms support for the North Atlantic Treaty nations because liberal Democratic senators who opposed ABM tended to be the same senators who wanted to reduce U.S. troop strength for NATO. Nixon was forced to deal publicly with ABM and NATO issues surrounding disarmament, even though arms reduction had not originally been part of his grand design.[66] Many of the same senators also opposed any expansion of the space program, and this complicated Nixon's problem in dealing with them because, like the president, they associated the ABM with the space program but, unlike him, they did not want to fund an aggressive program, whether civilian or military. Nixon's only public concession on the issue was to downgrade the "extensive ABM coverage," known as Sentinel under the Johnson administration, to a "reduced version" he called Safeguard—another major policy decision about which NASA was not consulted. The OST, PSAC, and NASA were simply out of the loop when it came to major foreign policy decisions that affected the space program.

During the spring and summer of 1969, Nixon dealt publicly and privately with NATO nations and constructed his gradually emerging détente policy—all the while battling U.S. senators over the ABM program—but conceding to their frugal perspectives on the space program. The president's handwritten comments and memoranda testify to his personal involvement in the domestic political fight over the ABM issue, but they do not indicate that he became personally involved in the House and Senate battles over the funding of other space efforts. The president persuaded most of the American public to accept his ABM proposals and, at the same time, to question increased spending for space. Of Senator Edmund Muskie's proposal to use the $6.6 billion slated for the ABM on hunger and poverty at home and abroad, Nixon wrote privately that it was "unbelievable nonsense from a national leader!" When he read that the former astronaut and then Ohio Senator John Glenn had called the ABM a "false hope" because "no one knows if it works," the president sarcastically asked: "Did he know the first space shot would absolutely work?" Obviously, Glenn's criticism did not encourage the president to accept NASA's requests for increased funding while the ABM debate raged during the spring and summer of 1969.[67]

In this political battle over the ABM system, Nixon and his staff never reconciled the potential contradictions inherent in Nixon's stance, namely, competition with the Soviet Union over the two countries' respective ABM systems and international cooperation with the Soviets in space. The administration's views and actions conveyed clearly that the space program was but one of many complicated issues it dealt with in its first months in office and that space took a backseat to most other issues. Fletcher's memoranda indicate that while Kissinger, along with his aides Al Haig and Jack Walsh, supported the continuation of some kind of human spaceflight, there was a "lack of advocacy in the Executive Office," except in the general area of international cooperation, because "they have been so busy" with other policy areas.[68] Fletcher and George M. Low, like Paine and Low previously, placed too much confidence in presidential advisors who did not deliver.

Given Nixon's basic foreign policy tenets, it should come as no surprise that from the moment he became president he and Kissinger, his National Security Advisor, urged NASA to pursue international cooperative opportunities, which the agency, some of its corporate clients, and some congressmen often found difficult to accept for ideological and commercial reasons.[69] Nonetheless, internal White House memoranda in the early 1970s emphasize that NASA was following rather than leading the way toward international space cooperation. A segment of one such memorandum stated: "NASA has been proceeding in this area with the understanding that it is responding to the President's policy," while another described Paine as "alluding repeatedly to what he described as [the president's] views" when encouraging "international cooperation in space."[70]

Less than two weeks after his inauguration, Nixon requested a "summary of European space activities . . . appropriate . . . to discuss with the Europeans." Paine sent him a six-page confidential response, emphasizing "additional ways in which you might express your personal interest in space cooperation." Paine indicated that the half dozen European nations developing their own space programs all feared that the United States would "impose its will on the direction of future West European space activities." Paine also kept Nixon informed about the results of his own trips abroad on behalf of space cooperation. The administration considered the issue of cooperation on space issues by Western nations in the post-Apollo era so important that in 1970 the National Security Council produced Decision Memorandum No. 72 creating an Ad Hoc Interagency Group headed by Arnold W. Frutkin to coordinate space cooperation.[71]

In retrospect, several important cooperative actions in space were completed during Nixon's administration. The first was the International Telecommunications Satellite Organization (Intelsat), which went into operation in December 1972. Although a global communications consortium had been formed in 1964 after the launching of the first Telestar satellites, ratification by 54 of the 83 Intelsat nation members took eight years of complicated negotiations, the most energetic of which came during Nixon's first term.[72] A second was the cooperative effort with the predecessor organizations of the European Space Agency to build *Spacelab,* a sophisticated laboratory that fit into the shuttle's payload bay and was flown several times in the 1980s.[73] However, the centerpiece of Nixon's space diplomacy, which emphasized international cooperation, became the much publicized Apollo-Soyuz Test Project (ASTP) that both Paine and Fletcher pursued at Nixon's insistence.

There was no substantive response from the Soviets about cooperation with the United States on space until the end of the 1969 and even then it was clear that the USSR was only interested in coordinating planetary "goals" and in exchanging the "results of *unmanned* planetary investigations." An interim step in this process resulted in the January 1971 NASA–Soviet Academy of Sciences agreement on space science and applications. But at that time Soviet policy still prevented "discussing future mission plans and experiments in advance." Finally, the ASTP became a reality in the summer of 1972 following Nixon's successful May summit meeting in Moscow, which included four cooperative agreements with Moscow on space, medicine, science, and technology. Although some secondary sources attribute this achievement to the efforts of either George M. Low, acting NASA administrator following Paine's resignation, or to his successor, Fletcher, it is unlikely that the necessary equipment for international rescue and the crew exchanges could have successfully taken place in 1975 if it had not been an important ingredient in Nixon's policy of détente with the Soviet Union.[74]

The Politics of Space in Congress: Disagreements and Investigations

Presidents Nixon, Ford, and Carter generally did not have to fight Congress over the space program because members of both houses fought each other over it at two levels. Sometimes, when they were not disagreeing over the merits of manned versus unmanned space projects and comparing their costs with domestic social programs, they were competing with one another to obtain space contracts for their respective states. The only notable bipartisan consensus that emerged in the 1970s occurred in opposition to Nixon's decision to abolish the Office of Science and Technology in 1973. This consensus was strengthened as Watergate-related events began to overtake the Nixon administration and when Vice President Ford implied he would support legislation to reinstate a science advisor in the White House. As president, Ford signed the 1976 legislation re-establishing the Office of Science and Technology Policy (OSTP), but with less clout than it had under Eisenhower and Kennedy. Then Ford angered some senators by appointing H. Guyford Stever, former head of the NSF, who had been accused in 1975 of mismanaging public funds in an NSF-funded project called "Man: A Course of Study" (MACOS). President Carter considered downgrading the office once again as part of his government reorganization plan, but he finally relented and appointed Frank Press as his OSTP director in 1977. However, the new president did not agree with the congressional interpretation of the 1976 act and finally overrode a portion of it in 1978 by issuing an executive order that transferred responsibility for preparing science policy reports back the National Science Foundation. By 1979, most of these differences over procedure between Carter and Congress had been ironed out, and the administration gave strong support to completing shuttle development. Thus, the decade ended on a note of cooperation between Congress and the White House.[75]

Of the three men who served as NASA administrators in the 1970s, Fletcher was more careful than either Paine or Robert Frosch in handling NASA contracts with the space industry, because these had been the source of bitter political controversy in Congress since the 1960s as individual congressmen fought each other over the awarding of lucrative space contracts. Such charges first became public in 1964 when it was discovered that NASA personnel at certain facilities were assisting contractors and universities in their regions to obtain procurement contracts. The most publicized investigation took place after the tragic fire that killed three

Apollo astronauts in January 1967. At one point during the investigation, Representative Olin E. ("Tiger") Teague (D-TX), normally one of the strongest congressional supporters of the space program, issued a broad indictment of NASA's exercise of quality control over North American Aviation, the Apollo capsule contractor. Although Webb left office highly praised by individual members of Congress, he left behind a history of contract problems that his successors could only ignore at their peril. As noted above, Paine resigned with similar charges of favoritism hanging over his head.[76]

Fletcher found that he too had problems concerning multimillion-dollar contract awards. He faced pressures from all of the normal sources—industry, congressional delegations, special interest groups, and so on—but his background as a Mormon from Utah brought a special challenge. Fletcher, for example, tried unsuccessfully to keep regional and religious politics out of the contract decision process for the solid rocket booster for the space shuttle. In February 1973, he replied to Senator Frank E. Moss (D-UT) about improprieties in an attempt by one of the senator's staff to influence the decision. Fletcher commented in a letter that he personally typed:

> I feel an obligation to respond to the numerous efforts made by your office of late to have this Agency, and, in particular myself, look with considerable favor at the placing of some of our business in your State. Not only would it be highly irregular to say the least, but might provoke the kinds of inquiries we are not prepared at this time to handle. . . .
>
> But the fact remains, Mr. Chairman, that my hands are tied for the time being. In my present position here at this particular Agency, it would be extremely difficult if not somewhat unethical for me to channel any more of our contracts towards your State without arousing further suspicion. . . .
>
> I should also like to call your attention to another matter along these same lines. One of your staff—I think you probably know who I am referring to—went so far as to insinuate sometime ago that I had a moral, if not a spiritual obligation to acquiese [sic] on some of [the] business issues previously raised by President [N. Eldon] Tanner [of the Mormon church]. This person voiced an unthinkable opinion to the effect that my Church membership took precedent [sic] over my Government responsibilities. Knowing that you share similar sentiments with me in the clear separation of Church and State, I would like to request that you take this unpleasant matter under advisement with the individual in question and explain just how serious and unconscionable [sic] those references were. In the meantime, I will see what else can be done for you.[77]

While Fletcher apparently combatted such religious overzealousness, he was not as successful in dealing with regional congressional power brokers. NASA agreed to award the solid rocket booster contract to Utah's Morton Thiokol, in no small part because of Moss's pressure on Fletcher.[78]

Clearly, the job of any NASA administrator was not an easy one when it came to making contract decisions. After a long drawn-out process, in 1975 the Utah-based Thiokol Chemical Company obtained a $1.59 billion space shuttle solid rocket motor contract, but only after such competing companies as Lockheed, United Technology Center, and Aerojet, members of the House and Senate, and governors representing states where those companies were headquartered unsuccessfully appealed the decision to the General Accounting Office (GAO).[79] The fact that NASA's procurement decisions were upheld in the face of contractor appeals did not alleviate the political controversy they caused in Congress at the time.

While Fletcher flatly denied having done anything improper, there seems little question that he was susceptible to pressure from western (especially Utah) politicians because of his local political allegiances and religious priorities.[80] Following the *Challenger* accident in January 1986, Fletcher came under especially intense criticism when the investigation into the accident found that it was a defective O-ring design in the Morton Thiokol boosters that caused the accident. Some believed, with some justification, that the shuttle in general and the Thiokol-built solid rocket boosters in particular had probably been Fletcher's poorest management decision and had fundamentally hurt the prospects of space exploration.[81] That blemish on his record was carefully scrutinized following the *Challenger* accident, and Fletcher was condemned as the official indirectly responsible for the tragedy because he set the course for the space shuttle and the Thiokol solid rocket boosters in the first place.[82] Even were there no evidence of bias—and Fletcher was officially exonerated of any wrongdoing—Thiokol would probably not even have been in serious contention for the shuttle's solid rocket booster contract had Fletcher not been at NASA and thereby susceptible to regional political and, in his case, unusual religious pressure.[83]

Aside from geographical and partisan disagreements over the awarding of NASA procurement contracts, Congress disagreed most during the Nixon administration over his decision to fund the space shuttle program and his insistence on ASTP as part of détente with the Soviet Union, in spite of the fact that Nixon held both up as cooperative projects that could save NASA money. The debates in 1972 over the shuttle probably represent the most partisan ones of the decade because of the pending presidential election. On the other hand, the Apollo-Soyuz Project produced in the early 1970s another kind of partisan debate because some of the strongest congressional defenders of the space program, such as Teague, were also adamant cold warriors who did not want to cooperate on anything with the Soviet Union.[84] That these two sets of congressional debates in the 1970s took place under Nixon's Republican administration should come as no surprise because both houses of the Congress were controlled by Democrats.

The shuttle had to compete with the priorities of Democrats (some of them potential presidential candidates) who wanted to fund domestic spending programs rather than any of Nixon's foreign policy endeavors. Moreover, these Democrats perceived space almost entirely as part of Nixon's geopolitical designs (even though the administration considered space a domestic budgetary issue). Consequently, Senators William Fulbright (D-AR), Edmund S. Muskie (D-ME), George S. McGovern (D-SD), Thomas F. Eagleton (D-MO), William Proxmire (D-WI), Jacob K. Javits (R-NY), and Walter F. Mondale (D-MN) all came out against funding for the shuttle program as did such members of the House as Bella Abzug (D-NY) and Les Aspin (D-WI). It should be noted, however, that Senators Hubert Humphrey (D-MN) and Henry M. Jackson (D-WA) supported the space shuttle program. While congressional opponents of the ASTP were not quite as prominent, with the possible exception of Teague, they were no less formidable.

Of all the partisan opponents of the shuttle program, Mondale pursued the issue with the most single-minded vigor. "Virtually all of the useful things we have gotten out of space: communications, earth surveillance, weather stations, navigation, the technology of instrumentation and miniaturization," he observed on the television program, *Issues and Answers,* in January 1972, "most of this has come about through unmanned instrumented [sic] flight." Mondale also introduced on the Senate floor a bill that would have killed funding for the shuttle program in FY 1973, but it was defeated on the floor by 21 to 61 on May 11, 1972. McGovern, the Democratic presidential candidate, went so far as to call the shuttle "Nixon's moondoggle" and an "enormous waste of money," and his first running mate, Thomas Eagleton, said that it would "deprive important social programs of much-needed revenue." Eagleton's argument was echoed by almost all the Democratic (and some Republican) opponents of funding for the shuttle.[85]

In addition to the two major partisan debates in the first half of the 1970s noted above, there were three other important attacks on space funding in 1974, 1975, and 1977, led by Representative Edward P. Boland

(D-MA), chair of the House Appropriations Subcommittee in charge of NASA and NSF programs. In 1974, Boland successfully opposed the "development of a large space telescope and deferred development of an experimental satellite to observe ocean characteristics (SEASAT)." Some funding for SEASAT was restored by the Senate. Then in 1975, Boland successfully delayed for one year the "active development" of the *Pioneer* satellite to explore the planet Venus, but again the Senate restored the funding for this mission. Finally, in 1977 Boland succeeded in getting the House to vote against funding to develop the *Galileo* probe to Jupiter.[86] All in all, however, Boland's efforts delayed rather than permanently cancelled the programs in the 1970s. By the end of the decade, he was still holding the line on NASA appropriations. However, Boland began working more cooperatively with Representative Don Fuqua (D-FL) when Fuqua succeeded Teague as head of the House Committee on Science and Technology.[87]

In summary, neither the three NASA administrators nor the three men occupying the White House in the 1970s experienced total defeat in Congress on any given space idea (with the exception of Paine's efforts to gain approval for a manned mission to Mars, although that never came to a vote in either chamber). Nonetheless, even Carter, the only Democrat of the presidential trio, and his NASA administrator, Frosch, the second Democrat to serve at the space agency in the 1970s, faced problems on Capitol Hill with even their modest space efforts from time to time. Since Carter was not an advocate for an expansive space program, and his vice president, Walter Mondale, did not have the same interest in it as either Johnson or Agnew, there was little impetus from the White House for greater NASA funding. Accordingly, Congress played a prominent role in shaping the space effort and continued its deceleration despite the entrenched power of several committees in both houses that traditionally favored strong funding for NASA. In retrospect, the most drastic decreases in NASA budgets, from various administrations' requests for NASA to appropriations by Congress occurred in FY 1959 (a drop of 20.6 percent), FY 1964 (down 10.7 percent), and FY 1968 (down 10 percent). These cuts *before 1970* were proportionately much larger than any that occurred in the following decade when the discrepancies between White House requests and congressional appropriations showed a positive rather than negative relationship, meaning that Congress usually appropriated more than the administration requested. Nonetheless, NASA's budget continued to decline from FY 1967–74, with the most precipitous declines taking place under Johnson, leveling off under Nixon, and beginning to rise slightly under Ford and Carter until in FY 1980 it reached $5.24 billion, almost equal to the previous peak appropriation in FY 1965 of $5.25 billion. During the same period, NASA staffing fell from a high of 34,000 in 1965 to 23,000 in 1980.[88]

Conclusion

A combination of political disputes in Congress, sometimes stimulated by White House policies and sometimes not, along with new and more demanding budgeting procedures, and the increased importance of White House personnel at the expense of science advisors or NASA administrators, combined to reduce funding for space in the decade of the 1970s. Without either a strong popular constituency to overcome these factors or effective leadership on the part of NASA to mobilize popular or congressional support as a counterbalance to the agency's decreasing importance inside the White House, or both, deceleration was inevitable. It did not, however, take place exclusively in the 1970s, nor did Nixon initiate it.

Curiously, in 1975 Art Buchwald, a humorist not known for his knowledge of space, touched on an important part of the problem in a conversation with Fletcher. In discussing why NASA manned flights and nonspace program applications had not been given more attention in the press since 1969, Buchwald unhesitatingly said it was because they were not controversial enough. "Webb was a very 'abrasive' guy [who] was always stirring up controversy," Buchwald told Fletcher. He recommended that the administrator "stir things up a bit." Fletcher, who was to head NASA for most of the decade sadly agreed but said that he wanted to avoid controversy after so much of it in the early part of the decade. In 1977 when the *Wall Street Journal* declared that Fletcher "had no flair for politics or publicity," Barry Goldwater (R-AZ) defended him, saying, "We have enough people heading agencies in this town with a flair for circus-style showmanship. It is a pleasure to have a man like Jim Fletcher who knows where he is going and what he is supposed to do and does it."[89]

Indeed, in the decade following the Moon landings, NASA seemed to have greater talent for attracting either the wrong kind of attention, especially early on, or no attention at all, particularly later in the 1970s. As a result, its budget, programs, and prestige suffered, and space policy took a backseat to a myriad of other concerns for those in the White House. NASA's programs were not simply the failure of presidential leadership—the so-called myth of the imperial presidency—or of NASA and congressional leadership but were related to larger questions facing the

United States in the 1970s. To a very real extent, the space agency was throughout the decade of the 1970s out of sync with U.S. political, cultural, and socioeconomic trends, and it is unlikely that affirmative leadership at any level could have overcome that. Rather, the broad themes of personnel, budgeting processes, foreign policy, and political factors combined with the leadership issue to bring about the deceleration of NASA in this period.

Notes

1. Both terms are used in John M. Logsdon, "The Space Program during the 1970s: An Analysis of Policymaking" (unpublished manuscript, 1974), pp. 4, 11, NASA Historical Reference Collection, NASA History Office, Washington, D.C. The circus atmosphere has not entirely disappeared from events surrounding the space program, as evidenced in 1993, when it was announced that the Agency's program called Commercial Experiment Transporter (Comet) had sold an advertisement and received $500,000 from Columbia Pictures to emblazon Arnold Schwarzenegger's name and that of his movie *Last Action Hero* on an unmanned NASA rocket launched in the spring of 1993. See *Washington Post,* Jan. 26, 1993, pp. B1, B8; *International Herald Tribune,* Mar. 4, 1993, p. 16, and Mar. 11, 1993, p. 16. All such advertisers in the future will be given a thirty- to sixty-second film clip of liftoffs to be used in television commercials.

2. For details about how Kennedy reversed Eisenhower's policies on space, especially in the area of policy and prestige, see Logsdon, *The Decision to Go to the Moon: Project Apollo and the National Interest* (Cambridge, Mass.: MIT Press, 1970). For a fine discussion of Eisenhower's measured response to *Sputnik* see Robert A. Divine, *The Sputnik Challenge: Eisenhower's Response to the Soviet Satellite* (New York: Oxford University Press, 1993).

3. It is often forgotten that in his farewell address to the nation, Eisenhower not only warned against a "military-industrial complex" but also against a "scientific-technological elite." Beginning with Eisenhower, however, no president until Reagan clearly indicated to the American people that militarization of the space program was inevitable given the satellite sytems needed by ICBMs and later by antisatellite systems and laser and particle beam weaponry. Instead, both the Vanguard satellite program and later Project Apollo were billed as "civilian" enterprises, which initially only military experts knew to be untrue because of the ready application of technology to military purposes. In contrast, Charles de Gaulle openly touted the military aspects of the French space program, while in the United States passive militarization of space had taken place for years. See James R. Killian Jr., *Sputnik, Scientists, and Eisenhower: A Memoir of the First Special Assistant to the President for Science and Technology* (Cambridge, Mass.: Harvard University Press, 1977); Paul B. Stares, *The Militarization of Space: U.S. Policy, 1945–1984* (Ithaca, N.Y.: Cornell University Press, 1985).

4. Apparently, Johnson borrowed this metaphor from his press secretary George Reedy, but the Soviets also used it. See Herbert L. Sawyer, "The Soviet Space Controversy, 1961–1963" (Ph.D. diss., Fletcher School of Law and Diplomacy, 1969).

5. Johnson, interview by Walter Cronkite, "Man on the Moon: The Epic Journey of *Apollo 11*," *CBS News*, July 21, 1969.

6. For several views of the decade, see Allen J. Matusow, *The Unraveling of America: A History of Liberalism in the 1960s* (New York: Harper and Row, 1984); Charles R. Morris, *A Time of Passion: America, 1960–1980* (New York: Harper and Row, 1984); Jonathan Schell, *The Time of Illusion* (New York: Knopf, 1976); and Tom Schactman, *Decade of Shocks: Dallas to Watergate, 1968–1974* (New York: Poseidon Press, 1983).

7. Gallup poll, Aug. 6, 1969. When asked if money should be set aside for a manned Mars landing, 53 percent of Americans opposed, while 39 percent favored this idea. A subsequent Harris Survey confirmed this when 47 percent said "no" to spending $4 billion a year for ten years to "explore the moon and other planets in outer space," while 44 percent said "yes." Cited in *Congressional Quarterly*, Feb. 13, 1970, p. 403; Louis Harris, *The Harris Survey Yearbook of Public Opinion, 1970* (New York: Louis Harris and Associates, 1971), pp. 83–84.

8. While Harris polls in 1966–73 showed that public confidence in science had dropped by 19 percent, confidence in most U.S. institutions, such as education, the military, and business, also fell in that same period. In fact, the counterculture group, which drew largely from middle- and upper-middle-class students, was not as distrustful of science as lower-class, less-educated Americans. See Amitai Etzioni and Clyde Nunn, "The Public Appreciation of Science in Contemporary America," *Daedalus* 103 (Summer 1974): 191–205.

9. James C. Fletcher, "Antitechnology Bias," *Air Force Magazine*, Sept. 1971, p. 53; Sylvia Doughty Fries, "Expertise against Politics: Technology as Ideology on Capitol Hill, 1966–1972," *Science, Technology, and Human Values* 8 (Spring 1983): 6–15; Fries, "The Ideology of Science during the Nixon Years: 1970–76," *Social Studies of Science* 14 (1983): 326–28, 337–38. See also Fletcher, interview by Robert Sherrod, Dec. 27, 1972; by Logsdon, Sept. 21, 1977; by Roger D. Launius, Sept. 19, 1991, NASA Historical Research Collection. All these interviews confirm Fletcher's belief that a strong antitechnology bias existed in the country in the early 1970s.

10. Some of this was reflective of the low priority that Nixon placed on the space program. The fact that Paine, a Democrat and an outsider to the administration, remained as NASA administrator until late 1970 indicates that Nixon had little interest in NASA and its program.

11. Richard M. Nixon, "The Research Gap: Crisis in American Science and Technology," Oct. 5, 1968, cited in Edward David to Ed Harper, with attachments quoting Nixon's campaign statements about research and development for civilian needs and space, Feb. 3, 1972, box 26, Papers and Other Historical Materials of Edward E[mil] David Jr., Office of Science and Technology Files (OSTF), Staff

Member and Office Files (SMOF), White House Central Files (WHCF), Nixon Presidential Materials (NPM), National Archives and Records Administration (NARA), Archives II, College Park, Md.; Joan Hoff, *Nixon Reconsidered* (New York: Basic Books, 1994), pp. 21–27, 115–21.

12. The first cost overruns became severe in 1962 and critical by 1963 as the original budget of $350 million for the Gemini Program reached $1 billion. See Roger E. Bilstein, *Orders of Magnitude: A History of NACA and NASA, 1915–1990* (Washington, D.C.: NASA [SP-4406], 1989), p. 69; Ken Hechler, *Toward the Endless Frontier: History of the [House] Committee on Science and Technology, 1959–79* (Washington D.C.: Government Printing Office, 1980), pp. 101–5.

13. Webb had apparently made a deal with President Kennedy to locate the Electronics Research Center (ERC) "in walking distance of both Harvard and MIT." The decision on the ERC's location was apparently made by three senior NASA administrators: Seamans, Dryden, and Webb. Although the facility became operational in 1965 and had 844 employees by 1969, the project was canceled by the Nixon administration on Dec. 29, 1969, according to Paine's statement, on the grounds that "NASA cannot afford to continue to invest broadly in electronics research as we have in the past." NASA then transferred the physical plant representing a $30 million investment and equipment worth $20 million to the Department of Transportation where it was renamed the Transportation Systems Center. The same pattern of controversial funding that was ultimately withdrawn after much time and expense plagued the Nuclear Engine for Rocket Vehicle Applications (NERVA) and the System for Nuclear Auxiliary Power (SNAP). See Hechler, *Toward the Endless Frontier,* pp. 219–31, 255–57, quotations are from pp. 229–30.

14. David to Peter M. Flanigan, n.d., box 35, David Papers, OSTF; George M. Low to Flanigan, Dec. 12, 1970, Flanigan Files, SMOF, WHCF, NPM, NARA.

15. Hoff, *Nixon Reconsidered,* pp. 44–49, 79–81, 92–94, 290.

16. "Dispute about Possibility of Life on Mars," *Aviation Week and Space Technology,* Aug. 18, 1969, p. 16; *Congressional Quarterly: On the Issues,* Feb. 26, 1972, p. 435.

17. Nixon, *The Memoirs of Richard Nixon* (New York: Grossett and Dunlap, 1978). There are only seven page references to the space program in this thousand-page memoir.

18. Ibid., pp. 428–30.

19. Paine, memorandum for the record concerning Jan. 22, 1970, about meeting the same day with Nixon, Flanigan, and Ehrlichman, White House President (WHP), Nixon, Correspondence with NASA, 1968–72, NASA Historical Reference Collection; David to Harper, Feb. 3, 1972, David Papers, OSTF, SMOF, WHCF, NPM, NARA.

20. Nixon, *Memoirs,* p. 428. In 1989, Scott E. Lewis prepared a detailed finding aid on all space resources in the Nixon Presidential Materials (NPM) at the National Archives and Records Administration (NARA). More than anything, it

reveals how little interest there was in NASA and space compared to other domestic issues addressed by the Nixon administration. One of Nixon's few personal requests about the space program illustrates its triviality in his agenda. In one of his numerous scribbled notations on news summaries, in November 1970, Nixon asked his science and technology advisor to find out about Alvin Toffler's book, *Future Shock*. This put the entire OST staff and prominent scientists on the President's Science Advisory Committee (PSAC) to work interpreting a pop culture book. Here is part of their reply: "Your inquiry . . . has stimulated a great deal of discussion and soul-searching among PSAC'ers and others in OST. An examination of several specific cases indicates that there is no evidence for the future shock phenomenon as described by Toffler. For example, our parents and grandparents who moved from the farms, first had railroads, electricity, autos and aircraft, and gave up fundamentalism under the impact of science and urbanization, experienced drastic change, yet they did not experience future shock. Stress or shock in individuals and societies appears to arise not from change itself but from a feeling of losing control over one's fate, which sometimes accompanies change. . . . It is indeed true that many people do feel that change is being imposed. The incomprehensibility of technology is a major source of this feeling. These conclusions do not suggest that the solution to the problem is simple. They do suggest that the situation can be remedied in part by providing broader, informed public participation in decision-making. . . . Techniques are appearing which can help in forecasting the effects of technology and in tailoring it to human purposes. A number of us in OST have thoughts about means for increasing public participation in this process. One of the most effective would bring decision making as it affects daily lives down to the state and local level in line [with?] the New Federalism. . . . People must have a feeling for both humanistic and scientific cultures. Not all people can have a foot in both camps, but they can be *fans* who know the score and can cheer or boo appropriately." See David to Nixon, Nov. 19, 1970; John R. Brown III to David, Dec. 14, 1970; David to Nixon, Feb. 22, 1971, all in David Papers, OSTF, SMOF, WHCF, NPM, NARA.

21. Paine, memorandum for the record, Jan. 22, 1970.

22. Logsdon, "Space Program during the 1970s," p. 3; Hoff, *Nixon Reconsidered*, pp. 83–84.

23. Paine to Nixon, Feb. 26, 1969, WHP, Nixon, 1968–72; PSAC, *The Space Program in the Post-Apollo Period* (Washington, D.C.: Government Printing Office, 1967), p. 14; Logsdon, "Space Program during the 1970s," p. 5 (Townes report quoted); Hoff, *Nixon Reconsidered*, p. 129.

24. Nixon to DuBridge, Feb. 8, 1969; Nixon to Agnew, David, and Paine, Feb. 13, 1969, David Papers, OSTF, SMOF, WHCF, NPM, NARA.

25. Agnew, quoted in *New York Times*, July 17, 1969, pp. 1, 22; Logsdon, "Space Program during the 1970s," p. 39, based on a Dec. 29, 1970, interview by Logsdon with Mayo. My research indicates that Agnew's lack of influence within Nixon's inner circle was evident from the very beginning on most issues with the

possible exception of Native American Indian policy. His lack of influence on space policy was confirmed in John D. Ehrlichman, interview by Logsdon, May 6, 1983, NASA Historical Reference Collection.

26. Ehrlichman, interview, pp. 26–27; Space Task Group (STG), *The Post-Apollo Space Program: Directions for the Future* (Washington, D.C.: Executive Office of the President, 1969), p. 3, NASA Historical Reference Collection.

27. After Nixon's inauguration in 1969, the initial momentum for change in most domestic and foreign affairs came from such free-thinking outsiders as Robert Finch, Daniel Patrick Moynihan, Henry Kissinger, and later John Connally. All of these men appealed to Nixon's preference for bold action and broad conceptualization. With the exception of Finch, none had been closely associated with him prior to his election as president. Political-broker insiders increasingly gained ascendancy over free-thinking outsiders within the first Nixon administration as his plans to reorganize became more corporate in nature and more central to his thinking. Gray-flannel types—many of whom he had known for many years, such as John Ehrlichman and H. R. Haldeman, the president's two closest aides; Leonard Garment, the liberal Democratic counterpart to Moynihan among Nixon's insider advisors; Arthur Burns, counselor to the president and later head of the Federal Reserve Board; Melvin Laird, secretary of defense; John Mitchell, attorney general; George Shultz, secretary of labor and later head of the OMB; and businessman Roy L. Ash, chair of the President's Council on Executive Reorganization—all played the role of political-broker insiders.

28. Richard P. Nathan, "The 'Administrative Presidency,'" *Public Interest* 44 (Summer 1976): 41–44; Nixon, *Memoirs*, pp. 337–42, 351–56, 764–70.

29. Ehrlichman, interview, p. 22. Nixon compared *Apollo 8* (the first circumlunar flight in December 1968) to two other "joyous things that had happened on [the same] day" in 1968: his daughter Julie's wedding and the release of the crew of the *Pueblo*. See Nixon, *Memoirs*, p. 329.

30. Logsdon, "Space Program during the 1970s," pp. 9, 45, 59–66. For Flanigan's reliance on Whitehead and the general adversarial relationship between White House staff and NASA, see Fletcher to Low, Nov. 5, 1971, Fletcher Papers, 1971, NASA Historial Reference Collection.

31. Paine to Nixon, Feb. 26, 1969, NASA Historical Reference Collection.

32. Fletcher to Low, Nov. 5, 1971, Fletcher Papers, 1971, NASA Historical Reference Collection; Flanigan to Nixon, action memorandum, with attachment of letter from Paine to Nixon dated July 31, 1970, Aug. 10, 1970, box 9, Flanigan Files, SMOF, WHCF, NPM, NARA. After Paine refused to stay on beyond September when asked by the Nixon administration, Flanigan urged Nixon to select Roger Lewis immediately as the new NASA administrator because he was everything Paine was not: "an excellent spokesman for NASA and the Administration . . . [and] a competent administrator." Lewis refused the job and the individuals suggested by Paine as his successor were not given serious consideration by the White House except George Bush, who was about to lose his bid for the Senate from

Texas. Bush apparently did not want the job anyway, thinking it not important enough, according to Flanigan's handwritten comments. At the time, Flanigan did not want to hold the position for Bush until the results of the November election were in because "by not acting [now] . . . the Administration looks indecisive. In addition," Flanigan told Nixon, "it looks as if NASA and the Space Program were not considered important to the Administration. Given the current condition of the space industry, this would be an unfortunate inference." Many have argued that Paine resigned over NASA budget cuts, but newspaper accounts indicate that he had personal financial concerns. He could not send his four teenage children to private schools for less than $15,000 and live in Washington on a salary of $42,500. This low salary, by corporate standards, was another reason it took Nixon five months to replace him (Richard D. Lyons, "Paine Quits Post in Space Agency," *New York Times*, July 29, 1970, pp. 1, 16; Thomas O'Toole, "Financial Needs Cited in NASA Head's Departure," *Washington Post*, July 30, 1970, p. A3).

33. Paine, memorandum for the record, Jan. 22, 1970; Fletcher to Low, Nov. 5, 1971, Joseph P. Allen to Fletcher (concerning NASA's relations with CIEP), Oct. 15, 1973, Fletcher to Allen, Nov. 7, 1973, Fletcher Papers, 1971, 1973, NASA Historical Reference Collection.

34. Low to Fletcher, Oct. 15, 1974, June 24, 1975, Fletcher Papers, 1974, 1975, NASA Historical Reference Collection. In particular, Richard Cheney and Donald Rumsfeld were thought to be "space buffs" and "friend[s] in court within the White House" under Ford. Fletcher and Low gave detailed personal presentations about the space program to both Presidents Ford and Carter—something they did not do while Nixon was in office. Nixon's PSAC discussed the possibility of reorganization as a way of making "full use of scientific and technological resources" and recommended that the president consult with the Ash Council about this. See David to Nixon, Feb. 22, 1971, box 27, David Papers. Nixon received two kinds of advice from the Ash Council. The first recommended horizontal *and* decentralized corporate designs and stressed the "values of economy and efficiency, span of control, policies-administration dichotomies, straight lines of authority, and accountability," and the second had to do with management by objectives (MBO). It was on the basis of the first functional reorganizational principle that the Ash Council recommended as early as October 1969 relocating the OST. The working papers of the Ash Council were among those opened by the National Archives in December 1986. Those that proved particularly important for this summary of government reorganization under Nixon can be found in boxes 71–72, WHCF, President's Advisory Council on Executive Organization (PACEO), NPM, NARA. I have also utilized a 428–page, bound, in-house summary of all the Ash Council's recommendations to Nixon, which was provided to me by John Whitaker, who served first as Nixon's cabinet secretary and later as undersecretary of the Interior Department; the summary is entitled "Memoranda of the President's Advisory Council on Executive Organization."

35. Hechler, *Toward the Endless Frontier*, pp. 67–73; Larry Berman, *The Office*

of *Management and Budget and the Presidency, 1921–1979* (Princeton, N.J.: Princeton University Press, 1979), pp. 85, 105–30; Berman, "The Office of Management and Budget that Almost Wasn't," *Political Science Quarterly* 92 (1977): 298; Charles Warren, "The Nixon Environmental Record: A Mixed Picture," in *Richard M. Nixon: Politician, President, Administrator,* ed. Leon Friedman and William F. Levantrosser (New York: Greenwood Press, 1991), pp. 198–99.

36. For a summary of scientific opinion at the 1973–74 hearings on Reorganization Plan No. 1, see Fries, "Ideology of Science," pp. 332, first quotation is from 330. On the OMB's power, see Glenn P. Hastedt, *American Foreign Policy: Past, Present, Future,* 2d ed. (Englewood Cliffs, N.J.: Prentice-Hall, 1991), pp. 116–17, second quotation is from p. 117. The first impoundment of allocated space funds occurred in 1972 when the OMB withheld $24 million specifically earmarked for retrofitting existing aircraft with quiet engines. See Hechler, *Toward the Endless Frontier,* p. 759.

37. Hechler, *Toward the Endless Frontier,* pp. 511–13; Fries, "Ideology of Science," pp. 328, 332–36, quotation is from p. 330. In this article, Fries also appears unaware of the influence of Watergate on testimony, of the importance of the Ash Council in all of Nixon's reorganizational proposals and New Federalist concepts, and of the growing power of individual presidential advisors.

38. Richard Rose, *Managing Presidential Objectives* (New York: Free Press, 1976), pp. 58–66.

39. Willis H. Shapley to Fletcher, Jan. 9, and Jan. 22, 1973, with attached notes; Low to Fletcher, Apr. 23, 1973; Fletcher to William Proxmire, Mar. 10, 1975; Larry J. Early, memorandum on Shapley's comments on the OMB and DOD, May 22, 1975; Fletcher to John E. Naugle, Apr. 13, 1976, all in Administrators Papers, James C. Fletcher, NASA Historical Reference Collection. See also Fletcher, interviews, 1972, 1977, 1991, NASA Historical Reference Collection.

40. George Mahon (D-TX), chair of the House Committee on Appropriations, to Fletcher, Oct. 13, 1976; Fletcher to Bill [William] Lilly, Oct. 19, 1976 (quotation), WHP, Carter, Correspondence, 1976, NASA Historical Reference Collection. This newspaper story reported that one of the benefits of zero-based budgeting was that NASA had decided to "discontinue the individual testing of space shuttle engine components prior to testing of complete system." In theory, ZBB required agencies to justify continuing activities from scratch—from a minimum or zero base— instead of taking current outlays for granted and focusing on increases, as NASA officials had routinely done.

41. Hechler, *Toward the Endless Frontier,* p. 207; Fletcher, interview by Sherrod, Dec. 27, 1972, NASA Historical Reference Collection; Bilstein, *Orders of Magnitude,* p. 58.

42. Hechler, *Toward the Endless Frontier,* pp. 191 (first quotation), 340 (second quotation). While it is easy to criticize Webb's lack of planning, one can sympathize with him at least a little. All of NASA's planning efforts throughout his leadership had been based on grandiloquent strategies first defined by Wernher von Braun in

the 1950s. Reining in those schemes would have been exceptionally difficult in the NASA institutional culture and almost impossible without understanding whether or not the budget would preclude development. Given the uncertainties regarding funding in the late 1960s as well as the scale of NASA planning, Webb was probably politically savvy not to submit a long-range plan. Instead, he concentrated on the Apollo mission and left the future to his successors. See Howard E. McCurdy, "Justifying Space Exploration: Lessons from History," *Space Times: The Journal of the American Astronautical Society,* Mar.–Apr. 1994, pp. 10–12.

43. Leonard Roberts, "A Study of Long-Range Planning in the National Aeronautics and Space Administration" (unpublished manuscript, Graduate School of Business, Stanford University, 1970), NASA Historical Reference Collection, pp. 10–32.

44. A summary of George Mueller's original "integrated" plan can be found in NASA, *America's Next Decade in Space: A Report for the Space Task Group* (Washington, D.C.: Government Printing Office, 1969), pp. 45–56. The account of Paine's response to both the PSG material and Mueller's plan is taken from Logsdon, "Space Program during the 1970s," pp. 28–30, which is based on Logsdon's interviews with the participants.

45. Paine to Nixon, Feb. 12, 1969; Paine, memorandum for the record, Jan. 22, 1970 (quotations), WHP, Nixon, 1968–72, NASA Historical Reference Collection.

46. As examples of this, see Paine to Agnew, Sept. 12, 1969; N. S. Stoer, Economics, Science, and Technology Division, BOB, to Mayo, BOB director, "Analysis of NASA Report to Vice President on Recent Interest/Reaction to the Space Program," Oct. 2, 1969; D. A. Derman, Economics, Science, and Technology Division, BOB, to Mayo, "Budget Appeals Session for NASA," Nov. 19, 1969; Paine to Mayo, Jan. 19, 1970, all in RG 51, series 69.1, box 51–78–32, NARA.

47. Russell C. Drew to DuBridge, with attachment indicating May 16 meeting with Senators Richard B. Russell, Warren G. Magnuson, and Margaret Chase Smith and Congressman Alphonzo Bell, May 15, 1969, box 35, David Papers.

48. Black to Agnew, Aug. 20, 1969; Seamans to Agnew, Aug. 4, 1969; Seamans to DuBridge, with attached Sept. 4 letter to Drew, Sept. 5, 1969, box 35, David Papers; Logsdon, "Space Program during the 1970s," pp. 38–39, esp. his interview with Seamans.

49. White House Press Conference, Sept. 16, 1969, box 35, David Papers; Logsdon, "Space Program during the 1970s," pp. 20–21, 44–48.

50. Logsdon, "Space Program during the 1970s," pp. 50, 53, NASA Historical Reference Collection; Mayo to Nixon, Sept. 25, 1969; Kenneth Cole to Ehrlichman and department heads, Sept. 30, 1969; DuBridge to Cole, Oct. 13, 1969, all three in box 35, David Papers. Mayo's criticisms were taken so seriously that Ehrlichman, Kissinger, Bryce Harlow, and Flanigan were all asked to review them.

51. All figures are from Logsdon, "Space Program during the 1970s," pp. 56–66, 69; Paine to Mayo, Nov. 18, 1969, NASA Historical Reference Collection.

52. BOB Staff Paper, "NASA Tentative Allowance—1971 Budget," Nov. 13, 1969, NASA Historical Reference Collection.

53. For Ford's and Carter's views, see the article from *Huntsville Times*, Sept. 14, 1976, and Ford campaign flyer—both attached to a note from Environmental Protection Agency at NASA to Fletcher and Lovelace, Nov. 1, 1976; White House, Office of the Press Secretary, "Fact Sheet: U.S. Civil Space Policy," WHP, Carter, Correspondence, 1976, NASA Historical Reference Collection; Richard D. Lyons, "Administration Discloses Plans for Use of Space Technology," *New York Times*, June 20, 1978, p. B4; Robert D. Toh, "New Space Policy Is Outlined," *Washington Post*, Oct. 12, 1978, p. A5.

54. Nick MacNeil to Stuart Eizenstat et al., Jan. 31, 1977; Presidential Directive/NSC-37, May 11, 1978; Presidential Directive/NSC-42, Oct. 10, 1978, copies in Jimmy Carter White House Files, NASA Historical Reference Collection; Lyons, "Administration Discloses"; Toh, "New Space Policy." See also Logsdon, "Opportunities for Policy Historians: The Evolution of the U.S. Civilian Space Program," in *A Spacefaring People*, ed. Alex Roland (Washington, D.C.: NASA [SP-4405], 1985), pp. 101–2, in which he discussed Carter's views on space issues in relation to Eisenhower's and Nixon's but missed the point that Nixon's were essentially related to his geopolitical theories, specifically détente with the Soviet Union.

55. For details about this contract controversy, see Anderson to Low, Sept. 17, 1970; Low to Anderson, Sept. 29, 1970; Spencer M. Beresford to Low, Sept. 29, 1970; Low to Flanigan, Sept. 30, 1970, all in box 9, Flanigan Files; editorial, *New York Times*, June 22, 1970, p. 44; Clark Mollenhoff, editorial, *Chicago Sun Times*, Sept. 13, 1970 (also inserted in the *Congressional Record*, Sept. 18, 1970), copy in NASA Historical Reference Collection.

56. Fletcher's attitude was especially evident in the first interview conducted with him by Robert Sherrod in 1972. The quotations are from Fletcher, interview by Logsdon, Sept. 21, 1977, pp. 30, 32.

57. Nixon, *Public Papers of the Presidents of the United States: Richard M. Nixon, 1969–1970* (Washington, D.C.: Government Printing Office, 1971), pp. 730, 252.

58. Low, memorandum for the record on his and Fletcher's Jan. 5, 1972, meeting with the president, Jan. 12, 1972, Fletcher Administrator Files.

59. Rice, quoted in interview by Logsdon, Nov. 13, 1975; Launius, "A Waning of Technocratic Faith: NASA and the Politics of the Space Shuttle Decision" (unpublished manuscript, 1991), pp. 12–15, NASA Historical Reference Collection; Senate Committee on Aeronautical and Space Sciences, Ralph E. Lapp testimony, Apr. 12, 1972, p. 7; Robert H. Hood to H. Dale Grubb, Apr. 13, 1972; Klaus P. Heiss and Oskar Morgenstern to Fletcher, Oct. 28, 1971, all in NASA Historical Reference Collection. For full testimony, see Senate Committee on Aeronautical and Space Sciences, *Hearings on S-3094*, 92d Cong., 2d sess., Mar. 22, 23, Apr. 12, 14, 1972 (Washington, D.C.: Government Printing Office, 1972), pp. 1051–109.

60. For specific reference to such activities, see Fletcher, notes, Aug. 5, 1971;

Fletcher to Shultz, Sept. 30, 1971; Fletcher to Low, Oct. 20, 1971, Nov. 5, 1971, and July 26, 1972; Fletcher to Weinberger, Nov. 3 and 4, 1971; Fletcher to William Morrill, assistant director of OMB, Oct. 2, 1972, all in Fletcher Administrator Files. For Nixon's views, see Fletcher, interview by Logsdon, Sept. 21, 1977, p. 26.

61. Weinberger to Nixon (via Shultz), Aug. 12, 1971, NPM, NARA.

62. Fletcher to Low, Aug. 24, 1971, Fletcher Administrator Files.

63. Weinberger, interview by Logsdon, Aug. 23, 1977; Ehrlichman, interview. See also Launius, "Waning of Technocratic Faith," pp. 15–20; Logsdon, "The Space Shuttle Decision: Technology and Political Choice," *Journal of Contemporary Business* 7 (Winter 1979): 13–30.

64. Franz Schurmann, *The Foreign Politics of Richard Nixon* (Berkeley, Calif.: Institute of International Studies, 1987), pp. 47–64, 84–90, 372–82. Schurmann makes a much more convincing case for Nixon's grand design than C. Warren Nutter does for *Kissinger's Grand Design* (Washington, D.C.: American Enterprise Institute for Public Policy Research, 1975).

65. Raymod L. Garthoff, *Détente and Confrontation: American-Soviet Relations from Nixon to Reagan* (Washington, D.C.: Brookings Institution, 1985), pp. 33–36, 47, quotation is from p. 33; Senate Committee on Foreign Relations, *Hearings on Détente*, 93d Cong., 2d sess., Aug.–Sept. 1974 (Washington, D.C.: Government Printing Office, 1974), pp. 239, 301 (quoting Dean Rusk and Kissinger); Richard W. Stevenson, *The Rise and Fall of Détente: Relaxation of Tension in U.S.-Soviet Relations* (Urbana: University of Illinois Press, 1985), pp. 6–11, 179–82, 188; Schurmann, *Foreign Politics of Richard Nixon*, pp. 80–81, 88. For the argument that détente reflected simply a continuation of George Kennan's ideas about containment, see John Lewis Gaddis, *Strategies of Containment: A Critical Appraisal of Postwar American National Security Policy* (New York: Oxford University Press, 1982), pp. 283–410.

66. Nixon, *Memoirs*, pp. 415–18; Schurmann, *Foreign Politics of Nixon*, p. 204.

67. Alexander Butterfield to Nixon, June 11, 1969; Butterfield, telephone conversation with Nixon, Aug. 1969; Bryce Harlow to Nixon, July 1, 1969, box 2, President's Handwriting (PN); box 30, 8, 13, Apr. 24, May 11, 1969, box 30, Annotated News Summaries (ANS), President's Office Files; Nixon to Gerbert Klein, Mar. 13, 1969; Nixon to Ehrlichman, Apr. 10, 1969, box 1, President's Personal Files, WHSF; J. Francis Lally to Nixon, June 10, 1969, box 64, General Foreign, WHCF, NPM, NARA.

68. Low to Fletcher, Aug. 12, 1971; Fletcher to Low, Nov. 5, 1971, Dec. 2 and 9, 1971; Fletcher to Low, Jan. 27, 1972; Fletcher to Joseph P. Allen, Nov. 7, 1973; Low to Fletcher, Oct. 16, 1974, all in Fletcher Administrator Files. In his August 1971 memorandum to Fletcher, Low wrote: "Kissinger stated that stopping manned space flight in the United States is entirely unsatisfactory, and [that] he would do everything in his power to prevent this happening." Yet Kissinger's memoirs pay

scant attention to matters involving space (in the first volume of over a thousand pages, there are only seven page references to the Apollo Program; the second contains even less), and the latest biography of him by Walter Isaacson includes only three brief references to Apollo flights—none of the mentions substantive.

69. Norman P. Neureiter to Flanigan, Mar. 31, 1971; Flanigan to Kissinger, July 23, 1971; Neureiter to David, July 26, 1971, all in box 35, David Papers; Philip H. Trezise to Donald B. Rice, Oct. 20, 1971; U. Alexis Johnson to Kissinger, Nov. 1, 1971; Kissinger to Flanigan, Nov. 1, 1971; Walsh to Kissinger, Oct. 21, 1971, Nov. 3 and 15, 1971; Kissinger to Johnson, [Nov. 1971]; Theodore L. Eliot Jr. to Kissinger, Nov. 12, 1971, all in box 9, Flanigan Files, SMOF, WHCF, NPM, NARA.

70. Russell C. Drew to David, Oct. 20, 1970, box 35, David Papers; William P. Rogers to Nixon, Apr. 29, 1972; Walsh to Fletcher, May 5, 1972; NASA's comments on Rogers's memorandum, Apr. 29, 1972, all in Fletcher Administrator Files. The degree to which Rogers was outside of Nixon's inner circle of foreign policy advisors can be seen in this memorandum in which he questioned Nixon's emphasis on cooperating with European nations in space.

71. Paine to Nixon, Feb. 12 and 26, 1969, Mar. 26, 1969, Nov. 7, 1969, Jan. 9, 1970, June 23, 1970, July 31, 1970, with attachments about NSDM 72; Kissinger to Paine, Feb. 10, 1970; Flanigan to Paine, July 2, 1970, all in WHP, Nixon, 1968–72, NASA Historical Reference Collection; Walsh to Herman Pollack, Feb. 18, 1972, Fletcher Administrator Files.

72. Intelsat communication, n.d., giving history of the consortium; Leonard H. Marks, "Report of the U.S. Delegation to the Plenipotentiary Conference on Definitive Arrangements for Intelsat, February 24–March 21, 1969," both in Intelsat Files, NASA Historical Reference Collection; Robert J. Samuelson, "Intelsat: Flying High, but Future Course Uncertain," *Science,* Apr. 4, 1969, pp. 56–57; "Remarks by Archer Nelson," *Congressional Record,* Nov. 25, 1969, p. E100029–30, clipping in NASA Historical Reference Collection; Katherine Johnson, "Japan, Australia Offer Intelsat Compromise," *Aviation Week and Space Technology,* Mar. 2, 1970, p. 20; "Intelsat Agreement Takes Shape," *Astronautics and Aeronautics,* Dec. 1970, pp. 15–16; [Lyndon B. Johnson], Executive Order 11277, Apr. 30, 1966, designating the Intelsat consortium as an international organization, *Weekly Compilation of Presidential Documents,* April 30, 1966, p. 600; Nixon's remarks to Intelsat Plenipotentiary Conference, May 21, 1971, *Weekly Compilation of Presidential Documents,* May 24, 1971, pp. 786–87; Richard D. Lyons, "Propriety of Space Shuttle Contract Is Questioned," *New York Times,* Aug. 21, 1971, p. M; *Space Daily,* Dec. 18, 1972, p. 209.

73. David Shapland and Michael Rycroft, *Spacelab: Research in Earth Orbit* (Cambridge: Cambridge University Press, 1984); Douglas R. Lord, *Spacelab: An International Success Story* (Washington, D.C.: NASA, 1987).

74. Paine to Nixon, Jan. 9, 1970; Fletcher to Nixon, July 28 and Nov. 22, 1972, all in WHP, Nixon, 1968–72, NASA Historical Reference Collection; Bilstein, *Orders of Magnitude,* p. 107; Hechler, *Toward the Endless Frontier,* pp. 208, 412.

75. Hechler, *Toward the Endless Frontier,* pp. 513–25, 605–54; Hans Mark, *The Space Station: A Personal Journey* (Durham, N.C.: Duke University Press, 1987), reviews policy during the Carter administration; "Focus for Science Advice, 12/15/76," Science and Technology, box 8, Al Stern Files, Carter Presidential Library, and "Report of Congressional Research Service to Stu Eizenstat, 2/15/77," Stu Eizenstat Files, Carter Presidential Library, both cited in Ron Fresne, "Science Advice in the Carter Administration: A Return to the Golden Years?" unpublished seminar paper, Georgia Institute of Technology, 1988.

76. Hechler, *Toward the Endless Frontier,* pp. 185–89, 194–206 (Teague is quoted on p. 196).

77. Fletcher to Moss, Feb. 23, 1973, NASA Historical Reference Collection.

78. The pressure of key congressmen on Fletcher to award the solid rocket booster contract to Morton Thiokol, especially by Utah Senator Moss, chair of the Committee on Aeronautical and Space Sciences, was omnipresent. See Fletcher to Moss, Feb. 22, 1972, and Jan. 12, 1973, NASA Historical Reference Collection.

79. For representative correspondence over the contested contract, see Calvin L. Rampton, Utah governor, to the Utah Delegation and Fletcher, Robert Curtin, and Dale Myers, June 24, 1971; Russell Long to Elmer B. Staats, GAO, Oct. 15, 1973; Staats to Long, Nov. 14, 1973; A. H. von der Esch, vice president of Lockheed, to GAO, Dec. 5, 1973; Rep. Dante B. Fascell (D-FL) to Staats, Dec. 21, 1973; GAO to Fletcher, May 9, 1974; Senators Russell B. Long, Bennett Johnson, F. James, and O. Eastland, to Fletcher, telegram, May 16, 1974; M. J. to Sen. Moss (outlining chronology of correspondence over contract award to Thiokol from Nov. 1973 to May 1974), May 22, 1974; Bob Allnutt to Moss, June 3, 5, and 12, 1974 (outlining NASA selection procedures), all in Frank E. Moss Papers, Special Collections, University of Utah. The final granting of the contract to Thiokol came in 1975. See Fletcher to Nelson A. Rockefeller, president of the Senate, Mar. 24, 1975, Fletcher Administrator Files.

80. Allnutt to Moss, "Comptroller General Meeting with William L. Walker, Governor of Mississippi, on Lockheed Protest of NASA Selection on Thiokol for Space Shuttle Solid Rocket Motor Contract," June 12, 1974, Moss Papers; *NASA Procurement: The 1973 Space Shuttle Solid Rocket Motor Contractor Selection* (Washington, D.C.: General Accounting Office, 1986); *Space Shuttle: Changes to the Solid Rocket Motor Contract TLSP: Report to Congressional Requestors* (Washington, D.C.: General Accounting Office, 1988); *Space Shuttle: Follow-up Evaluation of NASA's Solid Rocket Motor Procurement* (Washington, D.C.: General Accounting Office, 1989); *Space Shuttle: NASA's Procurement of Solid Rocket Booster Motors* (Washington, D.C.: General Accounting Office, 1986). Some writers have emphasized the connections of Fletcher, Morton Thiokol, the Mormon Church, and Utah interests in the awarding of the solid rocket booster contract. For instance, Malcom E. McConnell, *Challenger: A Major Malfunction* (Garden City, N.Y.: Doubleday, 1987), makes much of Fletcher's Utah connections in the award of the solid rocket booster contract to Utah-based Morton Thiokol. Anson

Shupe, *The Darker Side of Virtue: Corruption, Scandal and the Mormon Empire* (Buffalo, N.Y.: Prometheus Books, 1991), pp. 141–60, refers to the *Challenger* accident as a "Mormon conspiracy." More balanced in this discussion is Joseph J. Trento and Susan B. Trento, *Prescription for Disaster: From the Glory of Apollo to the Betrayal of the Shuttle* (New York: Crown Publishers, 1987).

81. See Alex Roland, "The Shuttle: Triumph or Turkey?," *Discover*, Nov. 1985, pp. 14–24; Roland, "The Shuttle's Uncertain Future," *Final Frontier*, Apr. 1988, pp. 24–27.

82. William H. Broad, "NASA Chief Might Not Take Part in Decisions on Booster Contracts," *New York Times*, Dec. 7, 1986; "NASA Chief Denies Bias in Choosing Rocket Maker in '73," *Philadelphia Enquirer*, Dec. 8, 1986; Broad, "Senator Urges New NASA Inquiry as Rocket Design Debate Goes On," *New York Times*, Dec. 8, 1986; Broad, "Inquiry on Head of NASA Is Ordered by Key Senator," *New York Times*, Dec. 19, 1986; John Wilford Noble, "Space Agency Backed by Hollings and Riegle," *New York Times*, Dec. 11, 1986, all from "NASA Current News folders," NASA Historical Reference Collection.

83. There are numerous examples of such patronage in the histories of NASA and other agencies. Patronage, which has been the grease of government machinery since the nation's beginning, raises important questions, but appointed officials have generally been practical people who dealt with the political process and struck compromises as necessary to further the goals of the effort as long as they were not illegal. Fletcher admitted that he might have been naive about this process in government and said he worked hard to ensure that "wheeling and dealing" did not violate his personal sense of ethics or the government's standards of conduct (Fletcher, interview by Roger D. Launius, Sept. 19, 1991).

84. Fletcher to Low, Jan. 24, 1974, Fletcher Administrator Files. For details of both debates, see Hechler, *Toward the Endless Frontier*, pp. 191, 269–305, 410, 412–24, 1010. Interestingly, in 1993 similar arguments were made by the new Clinton administration for cooperating with the Russians on a space station in order to save U.S. taxpayers' money, only to meet objections based not on anticommunist arguments but on Russia's economic instability in the post–cold war era. See *International Herald Tribune*, Apr. 8, 1993, p. 1, and Apr. 16, 1993, p. 3.

85. Mondale, quoted in *Congressional Quarterly*, Feb. 26, 1972, p. 436 (others quoted on p. 437). See also Mondale, Proxmire, and Javits to their Senate colleagues, May 9, 1972; Mondale to Anderson, Apr. 25, 1972; Fletcher to Mondale and Anderson, Apr. 25, 1972, all in Fletcher Administrator Files; Hechler, *Toward the Endless Frontier*, pp. 286–97 (McGovern quote on p. 289). NASA also did not endear itself to liberal Democrats in 1972 when it resisted complying with congressional legislation and executive orders aimed at eliminating sex discrimination. By 1974 the National Organization for Women declared both Nixon and Fletcher "unfit for their respective offices" and specifically called for the administrator's ouster because only 80 of 24,000 executive-grade employees at NASA were women.

See *Congressional Quarterly Almanac, 1974,* pp. 467–77, and *Current News,* May 29, 1974, p. 9 (NOW criticism).

86. Low to Fletcher, Jan. 2, 1974; Fletcher to Low, Jan. 7, 1974; Low to Boland, Oct. 10, 1974, all in Fletcher Administrator Files; "Budget Update," *Congressional Quarterly Almanac, 1974* (Washington, D.C.: Congressional Quarterly, 1975), pp. 92, 94, 474–77; House, Boland report to Committee of the Whole House on the State of the Union, 94th Cong., 1st sess., June 11, 1975, p. 56; *Congressional Record,* 94th Cong., 1st sess., July 19, 1977, pp. H23668–H23677.

87. Hechler, *Toward the Endless Frontier,* pp. 1002, 1009–10. This greater cooperation was achieved in part because Fuqua gave "subcommittee chairmen more autonomy than they . . . had before, without turning the committee over to them" (p. 1009). Fuqua also provided subcommittee heads with more staff assistance to accomplish their jobs.

88. "Budget Update," *Congressional Quarterly Almanac, 1974,* p. 474; *Statistical Abstract of the United States, 1992* (Washington, D.C.: Government Printing Office, 1992), p. 595. Of course, the $5.24 billion in funding in FY 1980 was worth far less than the $5.25 billion in FY 1965. In terms of 1991 constant dollars, the FY 1965 budget was over $21 billion while that for FY 1980 was slightly over $8 billion, according to the Advisory Committee, *Report on the Future of the U.S. Space Program* (Washington, D.C.: Government Printing Office, 1990), p. 4.

89. Fletcher to Low, Feb. 6, 1975, Fletcher Administrator Files; Jonathan Spivak, "Apathy I: NASA's Biggest Foe," *Wall Street Journal,* Feb. 25, 1977.

5

Politics Not Science: The U.S. Space Program in the Reagan and Bush Years

Lyn Ragsdale

With hundreds of American flags flying, Ronald Reagan stood before an audience of fifty thousand people at Edwards Air Force Base, California, on July 4, 1982, to welcome the return of the space shuttle *Columbia* and its crew. This marked the last test flight for the shuttle program and the start of what its sponsors hoped would begin a regular schedule of commercial and government flights. Comparing the space shuttle to the Yankee clippers of the early Republic, Reagan spoke of a "national space policy" that would "look aggressively to the future by demonstrating the potential of the shuttle and establishing a more permanent presence in space." The crowd cheered as Reagan declared, "Our freedom, independence, and national well-being will be tied to new achievements, new discoveries, and pushing back new frontiers" in space exploration.[1]

Reagan's remarks, coming after a decade of deflated expectations, cheered space enthusiasts. At last, it seemed, NASA had a friend in the White House. Not since John F. Kennedy boosted hopes with his 1961 decision to go to the Moon had a president seemed so clearly committed to a strong national space program. Pleadings by space advocates appeared to have reached sympathetic ears. In the years that followed, Presidents Ronald Reagan and George Bush seemed to confirm first impressions. Reagan approved NASA's long-desired plans for a space station, announced a commercial space launch policy, and rallied support for the bedeviled shuttle program after its first catastrophic accident (and he gave hope to space warriors with his 1983 commitment to the Strategic Defense Initiative [SDI]). Bush seemingly gave spirit and direction to the space program with his 1989 challenge to establish a lunar base and organize a human expedition to Mars.

Yet Reagan's remarks, certainly apropos to the symbolism and ceremony of the day, were nonetheless misleading. During the twelve years that the Congress and the Reagan and Bush administrations made decisions about space, *there was no national space policy.* Pronouncements emanated from congressional committees and White House councils, but they were often conflicting and incomplete in detail. Members of Congress and the White House staff issued many pronouncements about space but agreed on few.

Decisions about the space program did not constitute policy, if policy is defined by a reasonably well-thought-out plan of action to achieve a relatively well-defined goal to which both a congressional majority and the president agree. Instead, decisions were made to pursue projects "by the yard," taking small steps that advocates hoped would grow into large commitments.[2]

No one national direction for the U.S. space program characterized the Reagan and Bush years. The two presidents and Congress moved in independent directions regarding matters of space. Indeed, the term "the Reagan and Bush years" is only a descriptive convenience, for it obscures the fact that by the 1980s Congress had become an equal player with the presidency on space issues as well as on other affairs. Decisions on space that emerged were typically the result of compromises that left many matters unresolved. There was no one central, cohesive course set by either the president or any set of public officials.

A close examination of the events preceding Reagan's July 4 speech reveals these forces at work. Behind the patriotic staging and theatrical performance were political divisions of serious consequence. The occasion was used by members of the National Security Council staff to wrest control over national space policy from the president's science advisor and the cost-conscious budget director, David Stockman. NASA had encouraged the presidential visit in the hope of obtaining an early presidential commitment to a permanently occupied space station. To forward this long-sought initiative, NASA executives wanted Reagan to bless NASA's development work on the space shuttle as completed. Reagan did this with alarming alacrity, declaring the shuttle to be "fully operational." But instead of taking the next logical step, the most that the White House staff agreed to was a vague commitment to "a more permanent presence in space." Space advocates, nonetheless, put the best possible spin on the occasion in the belief that the details could be worked out later.[3] Incremental advances with details to follow proved to be the preferred method of policy making throughout the period.

Primary Policy and Ancillary Policy

In considering the space program during the Reagan and Bush years, it is important to draw a distinction between "primary policy" and "ancillary policy." Primary policy breaks with past decisions and perspectives to meet the nation's top priorities. Such policy establishes long-term goals and sets in motion organized efforts to achieve them. These include big ticket items that enjoy high agenda status—that is, they dominate public attention, public funds, and the deliberation of public officials.[4] Considerable media coverage, public opinion polls, legislative debate, and presidential communication are devoted to defining primary policy. Political parties judge their own success or failure based on their influence on these issues. During the Reagan administration, primary policy covered a triad of measures: budget cuts, tax cuts, and a large defense buildup. During the Bush administration, primary policy focused on the large budget deficit created in part by the incongruity between tax cuts and a massive defense initiative.

In contrast, ancillary policy does not set out to solve an identified national problem. Policy in its most concrete form consists of a plan of action designed to solve an identifiable problem. But ancillary policy is more apt to represent a continuing government commitment. It does not necessarily solve a current problem, even though the bureaucracy it maintains may have been set up for that purpose long ago. Ancillary policy has low agenda status; it receives only limited public attention, public funds, and efforts of public officials. Media coverage, public opinion polls, legislative debates, and presidential communication about the matter are sporadic at best. It is a matter of modest continuing importance that generates little interest among policy makers concerned about long-term directions or goals. With old ground already broken, policy makers develop new ground with many fewer grand strokes. Thus, primary policy is policy of innovation, while ancillary policy is policy of continuation.

In the 1960s, space policy was primary policy. In 1961, President Kennedy announced America's goal to land a man on the Moon by the end of the decade. That mission, framed as part of a race between the United States and the Soviet Union, gave the nascent space program a clear direction and purpose. It was a policy of innovation, instituted where nothing of a similar magnitude had existed before. Money flowed freely. Indeed, in the early going, NASA was hard-pressed to spend effectively all the money Congress appropriated.[5] There was general agreement on Capitol Hill, at the White House, and in the country on the singular lunar goal.

Although the goal held clear political implications, it was also a matter of scientific exploration—could men land on the Moon? What would be found there? In short, there was an innovative, well-defined national space policy.

Yet a great deal changed in the ensuing two decades. By the time that Reagan stood on the tarmac at Edwards Air Force Base, ancillary policy replaced primary policy in the space arena. Growing economic problems beginning in the 1970s, the declining value of space spectaculars as cold war showcases, and diminishing public interest in space after Project Apollo created a situation in which the space program was no longer a top national priority. The shuttle program was in mid-step. It had been an ongoing, but secondary, part of the government agenda for ten years. The program was not highly visible, but it never disappeared. Funds continued to be appropriated, but other major initiatives were on hold. In 1978, President Jimmy Carter had declined to undertake any new "high-challenge space engineering initiative."[6] Even with Reagan's stated interest in revitalizing the space program, it was not likely to return as a key national policy. Not only did the Reagan administration have other top goals, but there was no consensus between the White House and Capitol Hill or in the country as a whole about the direction space policy should take.

With ancillary policy, politics rather than the substance of an issue or problem drives decisions. This is because such policy exists only in the near term. It forgoes attention to coherent long-term strategy or planning. Absent a concern about long-term goals, the questions "What can we afford?" and "How can we sell it?" replace "What should we do?" The answers are marked by incrementalism and institutional conflict.

Incrementalism in Ancillary Policy

Ancillary policy entails incrementalism both in policy directions and budget decisions. As Howard E. McCurdy wrote, "In making an incremental decision, policy makers begin from an established base (generally defined as what the agency did in the previous year) and direct their attention not to the overall goals of the program but to incremental changes within it. By moving forward or backward from an established base, policy makers can change public policy without making final decisions about the long-term direction they are taking."[7]

Space advocates wanted the government to treat space as a primary policy, but this proved impossible in the years following the lunar landing. In 1969, advocates placed their full list of long-term objectives on the policy agenda. They asked the president and Congress to bless an orbiting space station around Earth, another around the Moon, a lunar base, an expedition to Mars, and a multipronged transportation system to reach all these objectives. Public support for an ambitious space program, however, had fallen to an all-time low. Only 14 percent of Americans polled favored increased spending on space, a figure that fell to 7 percent as the Moon landings multiplied.[8] President Nixon responded by approving one new initiative—the space shuttle. Sensing that the nature of policy review had changed, advocates decided to pursue their agenda incrementally. They would pursue their objectives one at a time. Once NASA completed development work on the space shuttle, NASA's administrator, James Beggs, asked for authority to begin work on what he called the "next logical step." With work on the space station underway, the next NASA administrator, James Fletcher, established an Office of Exploration to begin the preliminary studies necessary to propose a lunar base and expedition to Mars. Each new initiative during this period built on the previous base.[9]

During the 1980s, special study groups called periodically for a more comprehensive approach to space policy. George M. Low, head of the special transition team advising the incoming Reagan administration, pleaded for a long-range objective. Sally K. Ride and Thomas O. Paine each headed study groups that reviewed long-term goals for space.[10] In each case, however, their recommendations disappeared into the abyss of Washington policy debate. None of the reports resulted in the type of national commitment that had characterized the Apollo years: a single long-range objective on which the president, Congress, and the country agreed. Reviewing the future of the U.S. space program in 1990, a special advisory committee headed by Norman Augustine observed once again the "wide spectrum of perspectives . . . as to what its objectives should be."[11] Actual space endeavors proceeded incrementally.

Ancillary policy entails incremental funding. The established base, as McCurdy put it, is last year's budget, because the ongoing governmental commitment, which is at the heart of such policy, mandates continued resources. Budget shifts then occur at the margins—up and down; they amount to modest rather than sizable percentage changes. NASA's portion of the federal budget remained relatively steady during the Reagan and Bush years, despite assaults by budget cutters and efforts to implant new initiatives. Indeed, there has been little change in the funding of the space program relative to other portions of the federal budget since the 1970s. NASA's budget changed incrementally beginning in the early 1970s, as spending for Project Apollo tapered off. During the 1960s, NASA's share of the federal budget had skyrocketed from a minuscule 0.8 percent in fiscal year (FY) 1960 to a whopping 4.4 percent in FY 1966. By 1973, it had dropped to 1.3 percent. It hovered just below the 1.0 percent range during the next two decades. This contrast offers evidence of the space program's status as primary policy in the 1960s and its evolution toward a more ancillary policy in the later period.

A number of attempts to change NASA's share of federal expenditures—both up and down—occurred during the Reagan and Bush years. None managed to displace NASA's basic share of budget authority. In 1981, Budget Director Stockman presented a five-year plan that would have significantly reduced NASA's spending authority. Stockman argued that the agency's budget could be cut by a third without harming U.S. leadership in space. His plan was based on the widely held assumption that spending for space transportation would decline as the space shuttle became operational and brought in paying customers. Assuming that no new initiatives were approved, as Stockman hoped, NASA spending might be significantly reduced.[12]

Stockman's efforts were foiled in two ways. Spending on space transportation continued to climb throughout the decade, from $3 billion in FY 1982 to more than $4 billion in 1990. Plagued by technical difficulties, the shuttle never became a paying proposition. Before this became apparent, NASA won approval to start work on a permanently occupied space station. The endeavor was touted in part on the grounds that it would require only incremental increases in the NASA budget. Reagan approved the space station with the understanding that it could be funded through a modest 1.0 percent increase in NASA spending above inflationary increases.[13]

This modest increase for a multibillion dollar new project was based on the pretext that transportation spending would fall, opening up an expenditure wedge that could be applied to the development of the orbital facility. Employing this proposition, NASA expected space station funding to reach $2 billion by FY 1988—all new funds—without significantly altering the overall NASA budget.[14] In fact, space station funding did not break the $1 billion mark until FY 1990. Part of the continuing woes of the space station were driven by the realization that transportation funding did not fall. Full funding for the station therefore created more than incremental increases in the NASA budget, a situation widely resisted in Washington on the grounds that it allocated more than NASA's "fair share."

During the Bush administration, a similar attempt to increase NASA's budget also failed. In 1989, Bush approved NASA's Space Exploration

Initiative (SEI). The following January, he proposed a 24 percent increase in NASA funding for FY 1991, much of it to enrich existing efforts such as the space station program. Had the initiative gone forward, NASA's share of federal budget authority likely would have doubled from 1 to 2 percent in the years that followed. Such nonincremental increases were too much for even supporters of the space program on Capitol Hill and contributed to the early demise of the Space Exploration Initiative.

The only event to cause a nonincremental shift in the NASA budget was not part of an anticipated policy. Following the loss of the space shuttle *Challenger,* Congress approved a lump sum appropriation to build a new orbiter. In one year, NASA's budget jumped from nearly $8 to $11 billion— a 40 percent increase. In spite of assurances that this was a one-time occurrence, the higher funding level became part of NASA's base during the Bush administration. This allowed the president to press ahead with funding requests for approved projects and hold NASA's share of total budget authority at the post-*Challenger* level of 1 percent. As soon as Bush left office, however, the forces of incremental budgeting corrected this trend. Long-term cuts in the 1990s anticipated a return to pre-*Challenger* funding levels, as the budget gravitated back toward the incremental equilibrium point.[15]

The invisible hand of ancillary policy kept NASA spending in a fairly narrow range during the Reagan and Bush years. Space advocates dreamily recalled the peak funding of the 1960s and hoped for more. They lamented what they perceived to be lean funding imposed by economic conditions and political timidity, hoping that both would soon disappear. Foes of the program clamored for the diversion of space funds to social programs or deficit reduction. Neither group got its way. By its very nature, ancillary policy resists radical changes in spending levels, either up or down.

Although supporters of NASA are quick to lament the agency's "lean" budgets during the Reagan and Bush years, this predicament seems exaggerated (see table 5-1).[16] NASA certainly had less budgetary flexibility with the ancillary policy of the 1980s than with the primary policy of the 1960s. While budget incrementalism as a feature of ancillary policy indicates that funding for the program was not going to stop, neither was it going to soar.

Institutional Conflict

Institutional conflict—within and among Congress, the presidency, and NASA—impeded the extent to which space policy could be labeled "national" during the Reagan and Bush years. In a system of separate institu-

Table 5-1. NASA Budget, 1980–93 (Budget
Authority in Billions of Dollars)

Fiscal Year	Current Dollars	Constant 1993 Dollars	Percentage of Federal Budget
1980	5.4	10.3	0.8
1981	5.6	9.9	0.8
1982	6.2	10.1	0.8
1983	7.1	10.8	0.8
1984	7.5	10.8	0.8
1985	7.6	10.6	0.7
1986	7.8	10.6	0.7
1987	10.9	14.3	1.0
1988	9.1	11.3	0.8
1989	11.0	13.0	0.8
1990	12.3	14.0	0.9
1991	14.0	15.4	1.0
1992	14.3	14.9	1.0
1993	14.3	14.3	1.0

Source: Executive Office of the President, *Budget of the United States Government: Historical Tables,* FY 1996 (Washington, D.C.: Government Printing Office, 1995); NASA, Office of Comptroller, "Aerospace Price Deflator," May 1994.

tions sharing power through various checks and balances, institutional conflict is business as usual. In order to circumvent such conflict, officials in one or more institutions must be willing to invest political capital to raise public awareness and break with past practices. This is unlikely with ancillary policy.

The institutional conflict arises in part because the Congress, the presidency, and NASA are plural institutions.[17] In a plural institution, many hands go off in different directions to handle a multitude of tasks for which the organization has assumed responsibility. The plural organization is typically decentralized. A proliferation of units, many with equal status, clamor to make their viewpoint heard before final decisions are reached. There is little top-down control of offices and their efforts. Decision ambiguity results from decentralization. Decision makers are unclear about what is going on, what they want, how they will attain it, and who will be involved, *yet they make decisions nevertheless.* A plural institution is no small family business. Instead, a conglomerate of semifeudal, loosely allied offices with considerable independence from each other battle over incremental changes. Thus, the overall institution enjoys a life of its own, independent of any person or group within it.

Congress, known for the decentralization rooted in its committee system, was typically decentralized when considering NASA. Four committees oversaw NASA's work. The Senate Commerce, Science, and Transportation Committee with its Subcommittee on Science, Technology, and Space matched the House Science, Space, and Technology Committee with two relevant subcommittees: Investigations and Oversight, and Space Science and Applications. In addition, the House and Senate Appropriations Committees were keenly involved in space decisions. Both had subcommittees that handled the funding of Veterans Affairs, Housing and Urban Development, and Independent Agencies, including NASA.

Typically, authorizing committees play a major role in setting policy directions. During the 1980s, however, this was not true for NASA. The space authorizing committees received little respect from and had little power over the appropriations committees. Many authorizing committees develop informal understandings with their spending committee counterparts such that any authorization precedes appropriations. The space authorizing committees failed to reach such agreements with the appropriators.[18] So, the Appropriations Committee routinely acted on the space program before the authorizing committees finished their work. Often this meant that the Appropriations Committee was effectively making substantive authorization decisions without input from the authorizing committees. In addition, the appropriations committees would refuse to appropriate funds that had been previously authorized by the science committees. In part, this conflict between authorizers and appropriators occurred because the authorizing committees were consistently looked upon, in the words of the House committee's ranking Republican member F. James Sensenbrenner Jr. (R-WI), as "simply rubber-stamp[ing] NASA's wish list without prioritizing."[19] In addition, the House Science Committee often voted on its NASA authorization bill *after* the House Appropriations Committee had marked up its spending bill.[20] The Senate Commerce Committee acted even later than the House Science Committee, doubly diluting the impact of the authorizing committees.[21]

The delays were due partly to a battle between the House and Senate authorizing committees over whether NASA should receive a multiyear or a single-year authorization; the former was favored by the House committee, the latter by the Senate committee. The result of the battle mattered little in the Senate where, for part of the twelve-year period, Ernest F. "Fritz" Hollings (D-SC) was chair of the Commerce Committee and also a senior member on the Appropriations Committee. Thus, Hollings could exert his influence over the space program through the appropriations process without worrying about his own authorizing committee or its relations with the House committee. Disagreements and standoffs between the two authorizing committees dragged late into the year, year after year.[22] With the cart before the horse, the appropriations committees in both chambers largely set space policy through their funding decisions.

Several of the principal appropriators during the 97th–102d Congresses were also openly hostile to NASA funding, especially the space shuttle. From 1975 until the Republican party gained control of the Senate in 1981, William Proxmire (D-WI) served as the chair of the Senate Appropriations Subcommittee on Veterans Affairs-Housing and Urban Development-Independent Agencies (VA-HUD-IA). After the Senate returned to Democratic hands, he reclaimed the chairmanship in 1987 and 1988, after which he retired. During these years, Proxmire built a reputation as one of the most vocal critics of manned space ventures. He once bluntly described the space shuttle as "about the best example of a wasteful program I can think of."[23] On the House side, Proxmire's counterpart until 1988 was Edward Boland (D-MA). Boland, his successor Bob Traxler (D-MI), and Bill Green (R-NY), the ranking Republican on the House Appropriations Subcommittee on VA-HUD-IA, shared Proxmire's antipathy to big science projects with their correspondingly big budgets. While NASA had advocated funding its projects "by the yard," Traxler's measure was smaller. He felt that Congress should fund space projects "a foot at a time, and if all we can afford to put up there is two feet, then that's all we'll put." Traxler clashed with space supporters who believed that with tight budgets his logic would lead eventually to funding an "inch at a time" and soon no funding at all.[24]

Decision ambiguity was advanced by the nature of the arena within which NASA competed for funds. The subcommittee examining NASA spending also reviewed the departments of Housing and Urban Development and Veterans Affairs, as well as Independent Agencies, such as the Environmental Protection Agency. Thus, the appropriations subcommittees juggled federal spending requests among such strange bedfellows as space activities, low-income housing, veterans hospitals, and environmental cleanups. It was never clear in any given year how the appropriations subcommittee would respond to these competing requests. Thus, the funding climate was neither well suited to NASA's overall policy profile, nor well organized and cohesive in general.

Although Congress is typically viewed as a decentralized body prone to ambiguous decision making, it is much less common to view the presidency as a plural institution. Most people view the president as an individual. Yet, the presidency too is a plural institution characterized by organizational decentralization in the Executive Office of the President. It is also typified by decision ambiguity, as many of these units have overlapping and confusing jurisdiction on single policy issues. Presidents, especially when they are involved in ancillary policies, rely on these numerous offices to make decisions for them. The chief executives become involved in the decision process at its end, if at all. Thus, although there is more hierarchy in the White House than on Capitol Hill, presidents do not have (nor do they often wish to have) full control over all decisions made on their behalf.

The presidency acted as a plural institution in its decisions on the space program during the Reagan and Bush administrations. Decision making was decentralized to several units within the presidency that had jurisdiction over space. Their interests often conflicted, promoting issue ambiguity. Since the late 1960s, the Office of Management and Budget (OMB) had played a key role in the presidential review of national space issues. During the Nixon administration, the OMB had conducted the primary review work on the space shuttle decision. During the Reagan years, it was expressly interested in trimming the space budget. During the Bush years, the OMB was more inclined to enhance the space budget. The Office of Science and Technology Policy (OSTP), headed by the president's science advisor, also laid claim to space as science policy. Its influence was mercurial, dipping so low in the 1970s that the office was abolished briefly under Nixon, only to return under Ford. During the Reagan period, the science advisor favored unmanned exploration of space as faster and more efficient than manned flight.[25] The National Security Council (NSC) also had a major interest in space, preferring military space priorities to civilian ones. The military regularly spent more on space activities (primarily surveillance) than NASA did, and the NSC's position reflected this.

Finally, NASA itself is a plural institution. Its fifteen field centers are highly decentralized, operating more like rival universities than members of the same organization. Moreover, relations between field centers and NASA headquarters are typically strained, with field officials often suspicious of the role that the headquarters' officers seek to play.[26] Rivalry is especially intense among the field centers with responsibility for the human spaceflight program: the Johnson Space Center in Houston, Texas, the Kennedy Space Center at Cape Canaveral, Florida, and the Marshall Space Flight Center in Huntsville, Alabama. One researcher on NASA management summarized the competition among the three space centers:

"Each center nourishes a conviction that it is the best of the lot. Each center is hard at work to make its own place strong and secure in whatever lies ahead for NASA. No center is willing to reveal its entire hand to other centers, or for that matter, to Headquarters."[27] As will be discussed in greater detail below, their divided responsibilities both created decision ambiguity and turf fights about who should do what and exacerbated problems, such as those leading up to the *Challenger* explosion in 1986.

The nature of ancillary policy—the modest ongoing commitment reflected in incremental budgets, the absence of a long-term perspective, and the presence of institutional conflict—fueled the central contradiction affecting NASA during the Reagan and Bush years. NASA was interested in pursuing "big science"—innovative, technologically sophisticated, megaprograms that would take years to develop, more years to run, and even more time to perfect. Yet within the framework of ancillary policy, big science had to be accomplished cheaply and incrementally; it had to match the vested interests of a few and not upset other interests of the many; it had to be achieved with public support but without national excitement. In short, big science had to be crammed into small policy. And politics, not science—big or small—dictated the direction of small policy.

Both Reagan and Bush sought to reinvigorate the space program but found their attempts to exercise presidential leadership hampered by conflicts both within the White House and between the White House and Capitol Hill. Although they sought to forward the space program, they were not able to push it out from the framework of ancillary policy that had been in place for years. NASA's three largest initiatives—the space shuttle, the space station, and the missions to the Moon and Mars—were all affected by the contradiction between big science and small policy at the heart of national space efforts in the 1980s and early 1990s.

Billions to Spare: The Space Shuttle

The 1980s began with NASA, a beleaguered agency, facing tough questions: What kind of space program did the United States need? The triumph of the Apollo Moon missions seemed a dim memory. Indeed, NASA officials worried that Apollo's success bred the seeds of NASA's destruction. Public interest in astronauts had waned even as its interest in fiscal restraint had peaked. In the early 1980s, Thomas O. Paine, NASA's administrator from 1969–70, reflected, "The American people . . . didn't give a damn. By then, hell, we had been to the Moon. What do you care if we fly another orbital flight or not? We know we can do it."[28]

NASA had spent the 1970s developing the space shuttle, a reusable spacecraft that mated an orbiter to an external fuel tank and two solid rocket boosters. After the success of the Moon landing in 1969, NASA officials had urged that "the next logical step" in the space program lay in the development of a space station and a human mission to Mars.[29] When the Nixon administration rejected these primary policy initiatives, Fletcher pushed for the space shuttle as a more politically feasible alternative. "The only way to go," he decided, "was some sort of shuttle."[30]

The shuttle set NASA on a course of ancillary policy in a number of ways. It continued the agency's emphasis on manned spaceflight without setting any ultimate destination. Absent the space station, the shuttle was like an airliner with no place to land; it had no destination, except to keep NASA's manned spaceflight program alive until further initiatives could be approved. Many in the agency felt that the shuttle program was necessary for the agency's future survival. Fletcher argued in a letter to Caspar Weinberger, then Nixon's deputy director of the OMB, that "the shuttle provides the capability for a continuing U.S. manned space flight program, a capability we believe to be essential."[31] Walter Mondale, a vocal shuttle critic as senator from Minnesota, issued a harsher judgment: "There was this whole empire of people left over from the Apollo program with nothing to do. And to sustain their efforts, they needed show business. And manned flight was the drama."[32] Thus, the shuttle program offered NASA more of the same: it preserved its manned space mission without the high cost of a space station or a mission to Mars.

In other words, the shuttle protected the future of the space station through incrementalism and ancillary policy. NASA officials viewed the shuttle and space station as inseparable. Although the station might not be the very next step, it could be an eventual step once the shuttle flew successfully. How, so the logic went, could the space shuttle be developed without it some day serving as transportation for the space station?

Finally, the shuttle itself fit within the framework of ancillary policy. Lacking the space station as an immediate destination for the spacecraft, NASA officials promised economical, routine spaceflight instead. They compared the shuttle to regularly scheduled air travel.[33] The routine aspects of the program made the government's commitment ongoing; its intended affordability suggested that the commitment would not have to be overwhelming. Although an independent report concluded that cost should not be the chief criterion "to justify [the shuttle's] desirability," NASA made cost-effectiveness the chief selling point for the shuttle program with a wary White House and Capitol Hill.[34] NASA contended that not only would the shuttle offer routine trips to space but its customers would repay the cost of its development and operation "with billions to spare."[35] According to NASA, the shuttle would be cheaper than expendable launch vehicles (ELVs). Remote-controlled rockets that delivered satellites and other payloads into space were unlike the orbiter and its solid-fuel boosters because the former were not designed for recovery and reuse. In this way, NASA touted the shuttle as a good investment.

With the first orbital flight of the shuttle *Columbia* on April 12, 1981, NASA embarked on a new era of human spaceflight, both technologically and politically. Unlike the spaceflights of the 1960s, the new shuttle technology was not guided by any fully developed consensual space policy. What was fundamentally missing was any consensus on the shuttle's primary purpose: it continued to be all things to all people. It would be the nation's primary launch system for all payloads large and small, not just a shuttle carrying humans and material to an Earth-orbiting space station. Not only would the shuttle garner the economies of reusability, it would be cheaper to fly than any existing or future expendable launch vehicle. Not only would it provide routine access to space, it would serve as an orbital laboratory until more permanent space facilities could be established. While these promises made the shuttle attractive to policy makers, they created conflicting objectives that bedeviled engineers. Had Project Apollo sought to satisfy so many conflicting objectives, Americans would have never reached the Moon.

The conflicting objectives sought through the space shuttle were evident in several of Reagan's announcements on space. During 1981, an interagency work group within the administration, headed by the president's science advisor, conducted a full-scale review of space policy. Its product was Reagan's "national space policy," unveiled at Edwards Air Force Base on July 4, 1982. This defined the Space Transportation System (STS), the core of which was the space shuttle as the "primary" although not the exclusive "space launch system for both United States national security and civil government missions."[36] The policy reiterated a memorandum of understanding between NASA and the Department of Defense (DOD), worked out in January 1977.[37] The Reagan policy also spelled out twin priorities that the STS be both fully operational and cost-effective in providing routine access to space.

In a different vein, the president also encouraged private sector investment and involvement in space activities. By 1983, this encouragement took the form of a presidential pronouncement on the commercialization of space. In May, Reagan backed the development of expendable launch vehicles (ELVs).[38] Such ELVs competed directly with the space shuttle because they were unmanned, did not have to be returned to Earth, and were less expensive than shuttle missions. In August, Reagan met with corporate executives who informed him about the benefits of commercial space ventures that needed the shuttle and, most especially, the space station.[39] The businesspeople felt that an enlarged space industry would provide opportunities for manufacturing pharmeceuticals, computer chips, and metal alloys in space. The group told Reagan that presidential leadership was needed to stimulate the commercialization of space, which the president did. Thus, Reagan's push to reinvigorate the space program took the shuttle in numerous directions that were difficult to satisfy both politically and technologically. Throughout the 1980s, the shuttle program had many missions thrust upon it and had other ventures, such as commercial space endeavors, competing with it.

This mix of goals in a climate of ancillary policy made it difficult for NASA to build a shuttle fleet that was both reliable and cheap. As initially conceived, NASA projected the cost of developing the shuttle program at $8 billion. Each new orbiter would cost $250 million to build. There would be a fleet of five orbiters that together could make upward of fifty flights per year, with the first test flight scheduled for early 1978.[40] The shuttle was designed to transport approximately 32.5 tons of cargo into near-Earth orbit, that is, 100–217 nautical miles (115–250 statute miles) above Earth. It could also accommodate a flight crew of up to ten people, although a crew of seven would be more common, for a basic space mission of seven days. During a return to Earth, the orbiter was designed so that it had a cross-range maneuvering capability of 1,100 nautical miles (1,265 statute miles) to meet requirements for liftoff and landing at the same location after only one orbit. This capability satisfied the DOD's need for the shuttle to place in orbit and retrieve reconnaissance satellites.

By the time Reagan was inaugurated in 1981, the shuttle program had overrun its budget, costing $12.6 billion to develop (adjusted for inflation). Its development had taken three years longer than anticipated. The individual orbiters carried only 24 tons, cost $580 million to procure, and were expected to last only fifty missions. The cost per flight, initially estimated at $10 million, exceeded $94 million; the cost per payload pound, estimated at $100, reached $1,700. One of the most frequently cited cost-effectiveness studies used to defend the shuttle projected 580 missions during the first twelve years of operation; in fact, NASA flew the shuttle less than seventy times in 1983–94. The agency had promised that shuttle flights would be frequent, cheap, and manned. Instead, they were occasional, expensive, and manned. Conflicting objectives ensured that few if any goals were met.[41]

Big science is by definition expensive. NASA wanted to present the shuttle as big science made affordable. During tight budget times, the OMB wanted science made cheap. Congress wanted it to pay for itself by charging commercial and military users. And NASA tried to satisfy all. "We had to argue that it was cheaper," a NASA official noted. "It would be cheaper than the expendable launch vehicles. It would be better than all the expendable launch vehicles."[42] Thus, NASA was forced to offer the Congress and OMB "buy-in numbers," that is, projected costs that were low, foot-in-the-door estimates of programs with increases added incrementally. The estimates were not fictions but part of an effective political strategy. As Max Faget, one of NASA's premier engineers, remarked, "If you don't quote a low cost, you ain't going to get it to begin with."[43] Representative Bob Traxler (D-MI), chair of the House Appropriations subcommittee that funded NASA's budget, also acknowledged this strategy: "Mega-science projects are intentionally sold to us with a low price tag, with the understanding that as the project begins to gain momentum, it gains friends within the Congress and outside in industry, and that those kind of supporters will be able to roll the high numbers."[44]

Although the low buy-in figures helped NASA sell the shuttle program, the agency also paid a price for the estimates. As the agency began developing the shuttle, it became clear that cost overruns were likely. Yet the agency pursued, as one observer put it, the "myth of the economic shuttle."[45] The economic shuttle depended on a tradeoff between development and operating costs to keep within the $8 billion ceiling. Yet the Congress and OMB insisted that NASA lower both. This was patently impossible, but NASA executives feared that they would lose the whole program if they pressed this point. To lower development costs, NASA abandoned a fully reusable shuttle with a manned booster in favor of a semireusable shuttle with an expendable fuel tank. To decrease development costs further, the agency eliminated some $500 million in safety and other tests.[46] Despite these attempts to decrease development costs, cost overruns and missed budget projections were common. These increasing development costs then had to be offset by projecting relatively low operating costs. One OMB assistant director commented: "What needed more attention and never got any more attention was a good careful scrubdown of the operating costs. The number[s] that NASA was carrying around [were] absurd."[47]

But the cost overruns were more than the result of an external political strategy coming back to haunt the agency. They were also the result of internal practices, which included, according to auditors in NASA's Inspector General Office, the absence of competitive bidding, the failure to negotiate price agreements before work began, impractical deadlines, design changes in the middle of construction, and building parts before their design was completed and tested. A series of General Accounting Office (GAO) reports summarized the problem, deeming "generally ineffective" NASA's systems to track and correct management, financial, equipment, and property components. Most notably, the GAO observed NASA's inability to monitor adequately the work of thousands of contractors paid billions of dollars to manufacture and maintain the shuttle.[48]

The myth of the economic shuttle not only fostered unrealistically low estimates for the cost of development and operation but also established an unrealistically high launch rate. To defend the cost-effectiveness of the shuttle, NASA promised frequent flights. This would spread the fixed costs over a large number of flights, thereby reducing the cost per mission. An economic consulting firm advised NASA that the break-even point for cost-effectiveness required at least 25–30 flights per year. Originally hoping to fly the shuttle 48 times per year, NASA lowered this schedule to 25 missions per year as experience accumulated, although there were actually no more than nine flights in any one year.[49] Senior astronauts warned that more than nine flights per year would jeopardize the safety of the program, given the long time needed between flights for maintenance and preparation.[50]

Much of Reagan's attention to shuttle policy during the 1980s concerned the search for a sufficient number of missions to justify the program's existence. When Reagan was inaugurated, he was bullish on the space program and gave top priority to making the space shuttle operational and then utilizing it frequently and efficiently. Several circumstances defied the expectation of a fast launch schedule; some of the most apparent included competition from abroad, less than successful space commercialization efforts, and DOD requirements. The European Space Agency's (ESA) Ariane expendable launch system attracted well over $1 billion of business. NASA found it difficult for the shuttle to be competitive, because its payload prices significantly exceeded those of Ariane.[51]

The Reagan administration's efforts to boost the potential of space commercialization were never realized. Business executives had been overly optimistic about the markets for products developed in space, and the administration had been overly optimistic about the affordability and frequency of shuttle flights needed to take raw materials to the space station. So space commercialization efforts, which would have linked the shuttle to the space station and provided a strong rationale for both, were largely unsuccessful.

Even before the *Challenger* explosion, the DOD began to chafe at the notion of placing all its payloads on a NASA-controlled launch vehicle. Early in 1984, Secretary of Defense Caspar Weinberger issued a directive for the development and use of an ELV, which became known as the Titan IV, to supplement the space shuttle. While the directive indicated that the shuttle remained the "primary" launch system for DOD, ELVs represented a "complementary capability."[52] Subsequently, the Congress approved an Air Force request for FY 1986 that permitted the Air Force to spend $2.1 billion to build ten new Titan IV rockets and modify thirteen existing Air Force intercontinental ballistics missiles for satellite launches. After *Challenger*, the Congress approved $2.5 billion for another twenty-five Air Force ELVs.[53] NASA's four shuttle orbiters, regardless of their launch schedules, could not handle the volume of Pentagon business. Thus, although there was significant pressure to accelerate the shuttle program's launch schedule, the timetable was not fast enough either to make it commercially viable or to meet the Pentagon's demands for military payloads.[54]

In spite of the high hopes that had attended *Columbia*'s first launch in 1981, the shuttle program provided neither inexpensive nor routine access to space. By January 1986, there had been only twenty-four shuttle flights, although in the 1970s NASA had projected more flights than that for each year. While the system was reusable, its complexity, coupled with the ever-present rigors of flight in an aerospace environment, meant that the turnaround time between flights was several months instead of several days. In addition, missions were delayed for all manner of problems associated with ensuring the safety and performance of such a complex system. Since the flight schedule did not meet expectations, and since it took thousands of work hours and expensive parts to keep the system performing satisfactorily, observers began to criticize NASA for failing to meet the cost-effectiveness expectations that had been used to gain the approval of the shuttle program ten years earlier. Critical analyses agreed that the shuttle had proven to be neither cheap nor reliable—both primary selling points—and that NASA should never have used those arguments in building a political consensus for the program.

In some respects, therefore, by 1985 the effort had been both an achievement and a misadventure. The shuttle program had been an engagingly ambitious effort that had developed an exceptionally sophisticated vehicle. As such, it had been an enormously successful program. At the same time, the shuttle was essentially a continuation of space spectaculars, like

Project Apollo, and its much-touted capabilities at an affordable cost had not been realized. It made far fewer flights and conducted far fewer scientific experiments than NASA had publicly predicted.[55]

Thus, the two central expectations of the program went unmet. The shuttle was neither cheaper nor did it fly more frequently than ELVs. NASA oversold the program, but the president and Congress did little to correct this. Members of Congress and White House officials gave little attention to the shuttle's long-run launch capability or the future of space transportation. The myth of the economic shuttle, then, was just that—a myth.

Ancillary policy pushed NASA to compromise the feasibility of the project in order to obtain the political support to implement it. One Reagan budget official remarked, "I think they [NASA] have allowed their political assessment of what they have to do to get support to interfere with their scientific judgment."[56] That interference ultimately created conditions that contributed to the explosion of the shuttle *Challenger*.

Rush to Launch: The *Challenger* Crisis, 1986

On January 28, 1986, the space shuttle *Challenger* disintegrated shortly after takeoff, killing all seven astronauts on board. The accident was caused by the failure of two synthetic rubber O-ring pressure seals located in a joint between segments of the right solid rocket booster. The leak allowed white-hot combustion gases to burn through the joint. The failed booster struck the huge liquid-fuel external tank and the orbiter broke apart in a mass of burning hydrogen. To supporters and opponents alike, it was evident in the months following this tragedy that the U.S. shuttle program was in disarray. The explosion grounded the three remaining shuttles, suspended shuttle operations for thirty-two months, and triggered a fervent debate about the future of the entire space program. What was less clear to observers was the extent to which the shuttle program was in disarray *before* the accident. As the political scientist Maureen Casamayou has asserted, the accident was "not the painful, but inevitable, consequence of complex state-of-the-art technologies." It was not a matter, simply, that "sooner or later accidents were bound to happen."[57] Although it is always easy in hindsight to assess what went wrong, NASA had pieces of evidence regarding the seal joint early enough to anticipate trouble.

Several investigations followed the accident, the most important being the presidentially mandated blue-ribbon commission chaired by William P. Rogers, a former secretary of state. The Rogers Commission, after some prodding by the Nobel Prize–winning scientist Richard P. Feynman, did a credible if not unimpeachable job of grappling with the technologically difficult issues associated with the *Challenger* accident. Its report confirmed that the explosion resulted from a poor engineering decision, an O-ring used to seal joints in the solid rocket booster that was susceptible to failure at low temperatures, introduced, innocently enough, years earlier.

The commission also criticized NASA's internal communication system, finding that the potential for O-ring failure had been understood by NASA engineers prior to the launch of *Challenger* and that the accident could have been avoided. Indeed, a warning sounded as early as 1977 when one of NASA's own engineers at Marshall Space Flight Center in Alabama wrote that the primary O-ring might leak and that the secondary O-ring might not seal at all because of joint rotation. Two other memos to senior managers in the Marshall shuttle organization followed, one in 1979 and the other in 1983.[58] In December 1982, the O-rings were designated a "Criticality 1" feature of the solid rocket booster design, denoting "a failure point—without back-up—that could cause a loss of life or vehicle if the component fails."[59] Actual flight data from the *Challenger* launch in 1984 confirmed that the primary O-ring was susceptible to erosion.[60] In addition, a flight in April 1985 showed erosion of the secondary O-ring (meaning that the primary seal had failed) on a nozzle joint. Moreover, in 1985 Morton Thiokol, chief contractor for the booster, briefed officials at Marshall about cold temperatures weakening the O-rings.[61] Officials at Marshall dismissed the problem, and Thiokol did not pursue it. Finally, in 1985, Marshall officials informed NASA headquarters of the joint problem. The House Committee on Science and Technology also investigated the explosion and concluded, "The O-ring erosion history presented . . . at NASA headquarters in August 1985 was sufficiently detailed to require corrective action prior to the next flight. . . . [Yet] none of the participants . . . [from] NASA or Thiokol . . . recommended that the Shuttle be grounded until the problem with the seals was solved."[62]

Why then did NASA launch *Challenger*? In part, the launch went forward because Marshall engineers remained confident in the joint and sincerely did not expect it to fail. Detailed analysis of both documentary evidence and testimony show that NASA personnel involved in the O-ring question were convinced that the joints were safe and that there were numerous other problems—especially with the shuttle main engines—that consumed most of their attention. The fact that the seals had always operated successfully contributed to a sense that they would not cause a major accident. The catastrophic failure that occurred was a total shock to the Marshall Space Flight Center staff, made all the more painful by their perception that the Rogers Commission then used the center as a scapegoat.

Problems of institutional conflict within NASA also pushed the launch. The conflict was evident in the management of the shuttle's safety program, which involved a four-level review process to certify the readiness of all shuttle components. This process mirrored the shuttle program's overall organizational scheme.[63] Level 4 involved contractors who certified the flight readiness of shuttle parts. Level 3, which took place at Johnson, Kennedy, and Marshall Space Flight Centers, verified launch readiness at each center. Level 2, located at the Johnson Space Center, required certification of flight readiness to the manager of the entire shuttle program. Level 1 was at NASA headquarters in Washington, D.C., which held a "Flight Readiness Review" conference approximately two weeks before a launch.[64] Levels 3 and 4 were responsible for shuttle hardware; levels 1 and 2 were managerial and administrative.

Although this hierarchy was the primary channel of communication within the program, it was also the key to the communication failure in preparation for the *Challenger* launch in at least four instances. First, after the secondary O-ring failed in April 1985, the Marshall Solid Rocket Booster Project manager placed launch constraints on five subsequent shuttle flights, including the *Challenger* flight scheduled for January 1986. These constraints denoted that a problem, unless resolved, could potentially halt the mission. Still, on each occasion, the project manager waived the constraints prior to launch. Neither the constraints nor their waiver were known to levels 1 and 2. Second, on the eve of the launch, Marshall officials did not inform senior management at either level 2 or level 1 of the concerns expressed by Morton Thiokol engineers about launching in freezing temperatures. Finally, no mention was made of the O-ring problem at any level of the readiness review process for the 1986 *Challenger* launch.[65]

Amid the layers of the safety program, joint weakness was redefined as an acceptable risk. Morton Thiokol developed the rationale of a margin of "safe erosion" or "safe margin of erosion" in 1984, first on the primary O-ring and then later on the secondary O-ring.[66] Although safe erosion was a seeming contradiction, Marshall managers subsequently adopted the notion on flights beginning in April 1984. Lawrence Mulloy, Marshall's Solid Rocket Booster Project manager, testified at the House Committee on Science and Technology hearings: "I think we started down a road where we had a design deficiency. When we recognized that it had design deficiency we did not fix it. Then we continued to fly with it, and rationalized why it was safe, and eventually concluded and convinced ourselves that it was an acceptable risk. That was—when we started down the road, we started down the road to eventually having the inevitable accident."[67] "The irony, of course," Casamayou wrote, "is that, instead of becoming more alarmed as more flights showed evidence of erosion, the agency became more confident than ever that its predictions were sound."[68] In the words of one Rogers Commission member, NASA was "playing Russian roulette."[69] As long as the shuttle returned safely, none of these problems seemed daunting.

The flight of space shuttle *Challenger* began at 11:38 A.M. EST on January 28, 1986. It ended seventy-three seconds later. Hours after the shuttle exploded, President Reagan vowed, "We'll continue our quest in space. There will be more shuttle flights and more shuttle crews. . . . Nothing ends here."[70] The explosion of *Challenger* highlighted two very different messages about the U.S. space program. First, it pointed to the vulnerability of the space program, which people both within and outside NASA had not fully considered. Second, at the same time, because it served as a blow to American greatness in space and technology and because of the overwhelming outpouring of emotion after the tragedy, the accident promoted the continuation of the shuttle program and even its enhancement. As Reagan's remarks emphasized, the problems of O-rings and booster fuel segments were nothing that excellence and renewed commitment could not cure.

Indeed, Reagan's words renewed concerted presidential involvement in the space program. After the explosion, 1986 proved to be a busy year for the White House regarding space policy. In August 1986, the White House announced that a fourth space shuttle would be built to replace *Challenger*. In the same announcement, Reagan ended NASA's business of launching private satellites, leaving this to the various commercial companies that had filled the void by launching satellites during the time the shuttle program was grounded.[71] On November 6, 1986, Reagan vetoed the annual 1987 NASA reauthorization bill, rebuffing congressional efforts to create a National Space Council to oversee the space program but leaving funds to build a new orbiter safely intact in a separate spending bill. Reagan emphatically rejected the council as impeding his ability to "organiz[e] and manag[e] the Executive Office as I consider appropriate."[72] By year's end, Reagan asked former NASA administrator James C. Fletcher to return to the agency and assess the management issues facing the agency in light of the Rogers Commission report.

Yet, the certainty of the public announcements belied spirited controversy within the administration on exactly what to do after the *Challeng-er* disaster. In *Challenger's* aftermath, Reagan not only established the Rogers Commission but also instructed a subcabinet unit, created in 1982 and known as the Senior Interagency Group for Space or SIG (Space), to assess future options for the shuttle program. Originally established by Reagan to coordinate space issues, the SIG (Space) had a contentious past, reflecting the nature of the presidency as a plural institution. The eight-person advisory committee included the president's national security advisor as chair and representatives from NASA, the Army, Navy, Air Force, Central Intelligence Agency, the Arms Control and Disarmament Agency, and the Department of Commerce. Officials from OMB and the Office of Science and Technology Policy sat as observers.[73] Decision ambiguity and bureaucratic infighting were rampant. Moreover, Reagan's concerns about space after *Challenger* did not extend to strong directives or personal involvement in the SIG (Space)'s deliberations.

The group could not agree on a recommendation on the space shuttle program. It was locked in a dispute between those in favor of a new orbiter, including NASA and the Pentagon, and those who wanted to reduce space program costs, namely the OMB. After the shuttle explosion, the SIG (Space) took eight months to recommend a replacement orbiter but could not agree on a way to fund it.[74]

These disputes within the plural institution of the presidency at a time when clarity not conflict was of utmost importance led Congress in 1986 to call for the creation of the National Space Council; Reagan responded angrily with a pocket veto, chastising the Congress for telling him what to do.[75] The 1988 NASA reauthorization bill again called upon the president to create such a council, and this time Reagan, about to leave office, did not block the action.[76] In addition, the Congress, not the president, furnished the plan to finance the new replacement shuttle. In a display of institutional conflict at several levels, the Congress intervened after the dispute among White House officials in the SIG (Space) about the new shuttle's funding dragged on.[77] Lawmakers approved the transfer of $2.4 billion from the DOD to NASA to help pay for it.[78] Within this context of institutional conflict, presidential leadership can and did go only so far.

The shuttle program started anew in September 1988; the replacement shuttle *Endeavor* was launched in January 1992, six years after the disintegration of *Challenger*. Because the contractors knew how to build a shuttle, there were no cost overruns on *Endeavor*. A 1992 GAO audit reported that of twenty-nine programs initiated in 1977–91 with development costs over $200 million only *Endeavor* and three other projects came in under budget.[79] A more balanced space transportation program also began with the addition of a fleet of expendable launch vehicles. Despite these changes, NASA's own Advisory Committee on the Future of U.S. Space Programs concluded in 1990 that the shuttle program had still not emerged from "developmental status."[80] After the shuttle explosion, people inside and outside NASA agreed that the goal of cost-effectiveness and commercial competitiveness, which NASA, the presidency, and Congress had promoted some twenty years before, ought to be abandoned.[81]

Jobs Today, Jobs Tomorrow: The Space Station

In 1984, as an expression of its interest in reinvigorating the space program, the Reagan administration pledged support for the development of a permanently occupied space station. Echoing the lofty appeal made by President John F. Kennedy twenty-three years earlier, Reagan declared in his State of the Union message that "America has always been greatest when we dared to be great. We can reach for greatness again. We can follow our dreams to distant stars, living and working in space for peaceful, economic, and scientific gain. Tonight I am directing NASA to develop a permanently manned space station and to do it within a decade."[82] By the summer of that year, the Congress had approved a $156 million down payment for the space station in the FY 1985 NASA budget. Engineers and other NASA officials assembled at the Johnson Space Center to begin the preliminary design work.

Like the shuttle, the space station both bore the stamp of big science—a grand-scale, long-term technological project—and was pursued as ancillary policy. The space station was of secondary importance to the presidency and Congress. Despite Reagan's pledge of a reinvigorated space program, administration officials did not view the station as a top national priority. The project was of limited interest to most members of Congress, even to those from states where facilities would be built. More fundamentally, the space station was a product of incrementalism, both in substance and budget. In many ways, it was the space shuttle program repeated all over again.

Even more than the shuttle, the space station rested on budget incrementalism. As with the shuttle, NASA used the incrementalism to convince public officials that they could buy in at a relatively low price, for only a small portion of the project, and stretch purchases over a long time frame. A large space station would probably cost $13 billion (1984 dollars) just to design and assemble, to say nothing of the cost of transporting it to space and operating it once there.[83] Yet NASA originally estimated the station costs at $7.5–9 billion. As one NASA official observed, "I reached the scream level at about $9 billion," referring to the point at which he encountered resistance from politicians.[84] So NASA opted for an $8 billion price that would fit under its expected budget growth curve. The facility would be built piece by piece, for as much as the government could afford. Beggs, then NASA's administrator, told Congress in 1984, "The space station can be in place within a decade and for about $8 billion."[85] Edwin Meese, one of the station's top supporters in the Reagan administration, summarized the incremental strategy: "Let's get our foot in the water, so that we have a commitment and then we can worry about the long-range costs later."[86]

In identifying the objectives to be performed on the space station, NASA also used an ancillary approach. In the minds of most exploration advocates, a space station was a stepping-stone to the Moon and planets. No human spaceflight projects had been approved beyond low Earth orbit, however, so NASA was obliged to forgo this rationale in selling the program. To please as many potential users as possible, NASA's Space Station Task Force prepared a list of 107 missions that could be performed on the orbital facility during the 1990s. The space station would service satellites and act as a base for assembling other large structures in space. It would serve as a platform for observing the universe, as a research center, examining the effects of microgravity on materials and humans, as a transportation depot for satellites and people on journeys beyond, and as a factory in space, ushering in a new era of space commerce.

This was a tall order, especially for an $8 billion facility. As NASA engineers learned, all of these uses could not be accommodated in a single facility that cost only $8 billion to build. Yet NASA officials could not raise this issue in the White House councils examining the project. To do so would have required Reagan or someone on the White House staff to establish a primary goal for the overall space program, which could then guide the station's design. Lacking such a goal, NASA had to recruit as many allies as possible. In part to shield the engineering contradictions, John Hodge, the task force director, forbid the production of engineering drawings until the project was approved.

As design work proceeded, obstacles arose. A NASA investigating team concluded that a large number of parts would start to break down before the station was complete. The report concluded that the station would require 3,800 hours of maintenance annually, a heavy demand given the dangers of extravehicular activity.[87] Most critically, estimated costs continued to rise. The NASA field centers dealing with human spaceflight viewed the space station as a major source of funding. Every call for new estimates found the centers requesting more funds, which would be spent more for institutional support than for components in space. Policy makers began to calculate the added expenses of delivering the station to orbit and then operating it, especially as shuttle costs continued to soar. The station's cost, once set at $8 billion, was calculated at $38 billion through 1999.[88] While this included items originally omitted, such as transportation costs, and abandoned the pretext of calculating costs in 1984 dollars, the report nonetheless had a sobering effect on budget-conscious politicians.

The Congress and presidency could have used the new cost figures as a pretext for revisiting the entire space policy but did not do so. Instead, they too turned to incrementalism, the hallmark of ancillary policy. Both institutions doled out space station funds in small allowances, enough to keep the project alive but insufficient to meet the original deadlines. In NASA's FY 1991 budget, Congress officially ordered the agency to adopt an "incremental concept" by phasing in each portion of the station in a discrete and independent fashion. Funding for the station was capped for ninety days until NASA responded with a formal plan.[89] Both institutions also encouraged anxious NASA officials to reduce the scope of the space station—by the yard. NASA did so, shaving the functions of the facility until little more than an orbital laboratory remained. For both institutions, the space station was just attractive enough to survive but too expensive to fund fully.

Funding for the space station continued even though expectations of its role in the human exploration program remained murky. In her report on America's future in space, Sally K. Ride identified the space station as the operational hub for trips to the Moon and Mars.[90] This occurred even as administration officials lopped off station hardware that would make such capabilities possible. In the absence of a clear vision of station purpose, public officials turned their attention to jobs. The space station was a good domestic spending project that would orbit above Earth. By 1992, the project had spawned an estimated 75,000 jobs in 39 states, most in California, Alabama, Texas, and Maryland. Opponents and supporters of the station saw its importance for constituency interests. Representative David Obey (D-WI) stated, "There is no bigger pork item in the domestic budget than this item." The space station also served as a lightning rod for criticism regarding the high cost of big government programs in the midst of tight budgets. Senator William Cohen (R-ME) argued, "When we stand on this floor and argue day after day about the size of the budget deficit and then agree to fund programs of this magnitude, then I say there is no hope we will ever bring our budget deficit under control."[91] The opposition was never sufficiently well organized enough, however, to terminate the program. It was difficult to mobilize public opinion against the space program as the source of the federal deficit. And the scientific community remained divided on the project's merits.[92]

Space station supporters like Representative Tom DeLay (R-TX) responded, "Can you deprive your state and your constituents of this important source of jobs and revenue?"[93] Senator Barbara Mikulski (D-MD), the chair of the Senate Appropriations Subcommittee that handled NASA's budget, agreed with DeLay, "I truly believe that in space station *Freedom* we are going to generate jobs today and jobs tomorrow—jobs today in terms of the actual manufacturing of space station *Freedom*, but jobs tomorrow because of what we will learn."[94] The public works expectation was matched by the space station's incrementalism. The space station hardware could be developed and produced one piece at a time at various locations across the country.

Like the space shuttle, what was missing from the congressional debate on expectations about the space station was the ultimate purpose of the undertaking. Thomas Donahue, chair of the Space Science Board of the National Academy of Sciences, raised exactly this issue, "If the decision to build a space station is political and social, we have no problem with that. But don't call it a scientific program." He continued, "The Board sees no scientific need for this space station during the next twenty years."[95]

The lack of clear expectations about the role of the space station beyond job creation was evident in debates within the White House and Congress. When Beggs, NASA's administrator, initially proposed the space station to the Reagan administration, there was little consensus about its merits. Although most members of the Reagan administration favored a more ambitious space program, they did not necessarily see the station as part of that ambition. Officials in the OMB, especially Stockman, its director, argued against the station's excessive cost. Defense Secretary Weinberger wanted NASA to invest funds in the space shuttle, which the military had agreed to use as its primary launch system. The DOD saw little need for an orbiting facility, convinced that automated satellites would serve its needs for the foreseeable future. The president's science advisor wanted NASA to invest in more innovative missions, such as the exploration of Mars.[96] Nearly every one of the station missions NASA proposed, the assistant science director argued, could be better accomplished by satellites and the shuttle orbiter. In contrast, William Clark, Reagan's national security advisor, White House counselor Edwin Meese, and Meese's assistant Craig Fuller favored the station.

These viewpoints found their way into the SIG (Space), the advisory committee Reagan had assigned to coordinate space policy. Two SIG (Space) working groups examined the station proposal, producing mounds of paper but little agreement. After months of negotiation and intense disagreement, Meese and Fuller shifted space station considerations from the SIG (Space) to the agenda of the Cabinet Council on Commerce and Trade. Although the disagreements remained, NASA had more allies on this second subcabinet council. Reagan attended the December 1, 1983, meeting and heard NASA's presentation. With little consideration of the long-term consequences, Reagan decided to approve the endeavor. Generally, the president intervened only when his assistants could not agree.

Nonetheless, presidential blessing was insufficient to silence disagreements, especially as the program moved to Capitol Hill. Almost yearly, the project faced some new assault by people bent on altering either the space station configuration or its method of assembly. One of the most serious challenges occurred in 1991. For the first time in many years, the House Committee on Science and Technology reduced NASA funding in the authorization bill and did so prior to the markup by the House Appropriations Subcommittee. The authorizing committee trimmed the FY 1992 NASA budget by nearly $500 million, although it authorized full funding for the space station.[97] On May 2, the House passed the NASA authorization for $15.3 billion, $2.1 billion of which was for the space station.[98] Yet on May 15, the Appropriations Subcommittee called for the elimination of the space station, despite the earlier reauthorization bill. Traxler, the subcommittee chair, asserted, "We simply can no longer afford huge new projects with huge price tags while trying to maintain services that the American people expect to be provided."[99] On June 6, however, rebuking the subcommittee's action, the full House voted to continue funding for the space station by freezing every other space program and cutting public housing money for the poor. Funds for housing were restored after a lobbying campaign by the White House and NASA, which circulated among members a district-by-district breakdown of space station contracts, employment, and money spent.[100] House and Senate conferees ultimately agreed to continue the space project for one more year, allocating $2.03 billion for the station after cuts in other NASA activities.[101]

Ancillary policy suggests that a program might be cut but probably not abandoned. Ultimately, public works expectations and incremental inertia allowed the space station program to limp along. NASA spent the entire $8 billion originally estimated as the cost of the completed facility simply redesigning the station to meet congressional and White House demands. Although Reagan had boldly proposed a space station in 1984, he in fact had much less influence than the Congress on this space initiative. Indeed, the Congress continued to play a major role in this ancillary policy, doing what it does best—protecting local interests, offering jobs to constituents, and claiming credit for both.

During its two terms, the Reagan administration was active in matters of space. Reagan had announced a national space policy in 1982 that called for and hinted at a permanent place in space. In 1984, he called specifically for the development of the space station. In 1986, after *Challenger,* he called for construction of a fourth shuttle. And Reagan also balked at congressional efforts to organize and manage space policy through the National Space Council. But NASA programs, especially the space station, continued to face budget constraints and opposition.

1989: A Mission to the Moon and Mars?

On July 20, 1989, George Bush marked the twentieth anniversary of the first Apollo Moon landing by proposing an ambitious Space Exploration Initiative (SEI), which would return people to the Moon and mount a human expedition to Mars. Ten months later, speaking at Texas A&M University, Bush set a deadline for the pinnacle endeavor: a manned landing on Mars by 2019.[102] In the Moon-Mars Project, the Bush administration appeared to be offering a plan not just about big science but about biggest science. In doing so, it sought to bring the space program full circle back to the early 1960s and to primary policy.

For years, space advocates had been pressing for such a commitment. Flights to the planets appeared in NASA's 1959 long-range plan, in the recommendations of the 1969 Space Task Group, and in the 1986 report by the National Commission on Space. Throughout the late 1980s, following the approval of the Earth-orbiting space station, NASA had pressed for such a decision. In 1987, Fletcher had established an Office of Exploration to coordinate NASA's efforts to sell the Moon-Mars idea. This had resulted in a mild gesture by outgoing President Reagan, who in 1988 approved a national space policy that generally endorsed the notion that the United States should "expand human presence and activity beyond Earth orbit into the solar system."[103]

As with previous initiatives, space advocates viewed presidential blessing as the key to their endeavors. Mark Craig, one of the NASA officials in charge of the staff work that led up to Bush's pronouncement, called it "the paradigm of *Apollo*." Space advocates believed that the mission would begin once the president spoke in the appropriate way. "The man stands up, he makes a speech, he sets the date, and the money shows up at the door the next day," Craig remarked. Yet that was not what happened with Project Apollo, he observed, "and it is certainly not what happened here."[104] Nonetheless, many of NASA's efforts were directed at securing the appropriate presidential benediction.

The attempt began during the Bush transition when the leading candidate for the NASA administrator's post, Richard Truly, submitted a one-page white paper to the White House Chief of Staff John Sununu, suggesting that the administration set bold new national space goals. Sununu and Bush were receptive to the idea. The following month, Vice President Dan Quayle also raised the possibility of a new space initiative with Bush. The newly inaugurated president turned the matter over to the recently constituted National Aeronautics and Space Council (NASC), which Quayle headed. Bush had established the NASC by executive order in 1989 (after Reagan's pocket veto and congressional repassage). The council reflected the workings of a plural institution. The council's membership included the secretaries of state, treasury, defense, commerce, transportation, and energy, the directors of OMB and the CIA, the national security advisor, the president's chief of staff, and the NASA administrator.[105] Indeed, the group was more diverse than the SIG (Space), representing the proliferation of executive branch institutions with perceived interests in space.

Initially, the NASC staff requested an initiative that NASA could complete by the year 2000—most likely a return to the Moon. As discussions between NASA and the council ensued, the proposal grew more elaborate. If the purpose of a lunar base was to test technologies for the exploration of Mars, why not make that the long-term goal? The White House staff was receptive. The proposal that emerged recommended a lunar outpost followed by an expedition to Mars.

In the weeks that followed, Truly and Quayle briefed a cross-section of space policy decision makers. Members of Congress and their staffs, White House officials, scientists, and corporate leaders heard the ambitious plans. Remarkably, no one raised any serious objections. Conflicts that had divided policy makers during consideration of the space shuttle and space station programs did not reappear. Congressional representatives were astonished that the White House would consider such a plan during a period of fiscal restraint but did not object to the underlying idea. White

House officials who might have otherwise raised objections, such as the science advisor and budget director, were supportive.

NASA officials viewed the response as a rare confluence of support, the result of promoting space exploration to the point that it had become ingrained in American culture. In fact, a more troublesome force was at work. By its nature, primary policy requires a full commitment of resources necessary to accomplish the selected goal. Ancillary policy, on the other hand, is deliberately vague. People are asked to sign on to a general idea in the hope that they will not notice that their interpretations of it are irreconcilable.

In preparing the president's proposal, those doing the actual staff work attempted to stuff a primary policy into an ancillary box. NASA staff studies used to brief the vice president adroitly avoided any cost estimates. The study papers did not require NASA to reconstruct ongoing programs in any fundamental way. As a practical matter, the presidential declaration did little more than call for further staff studies. Bush directed, "The Space Council will report back to me as soon as possible with concrete recommendations to chart a new and continuing course to the Moon and Mars." This was in stark contrast to Kennedy's lunar proposal, which contained a half dozen specific proposals and the funding requests for them.[106]

In the following January 1990, planning for the initiative took a different turn. Instead of emphasizing the endeavor's low start-up costs, White House officials deliberately made the project look large. The president's budget request to Congress grouped together a number of supporting initiatives, including the National Aerospace Plane, a joint program to develop nuclear-fired electric-power generators in space, and the already approved Mars *Observer*. These were added to the relatively modest expenditures sought for more SEI staff studies, which brought the total SEI budget to $1.3 billion. This amounted to a nonincremental increase of 47 percent over FY 1990 levels. Bush himself stressed that his proposed NASA budget contained "the largest increase for any major agency of the government."[107] Although no clear financial plan was published while consideration of the initiative was underway, NASA officials anticipated that the endeavor would eventually require NASA's budget to more than double to about $30 billion per year, plus adjustments for inflation.[108]

The impression of presidential leadership in an era of ancillary policy did not work. Here was a clear presidential direction to the civil space program, nearly as definitive as Kennedy had asserted in 1961. Yet, in the early 1990s, the political climate of budget cuts, divided government, congressional parochial interests, and a faded Soviet space race precluded the successful exercise of presidential leadership. Merely because Bush said America would go to Mars did not make it happen. Prevailing ancillary policy directed the course of the space program and dictated the demise of the Moon-Mars Program. The initiative failed because the space program as a whole was considered a secondary priority by members of Congress who saw Bush's announcement as "just another space speech."[109]

Lawmakers objected to such an elaborate undertaking while existing projects like the space station were incomplete. The House Appropriations Subcommittee responsible for NASA's budget decided against funding the Moon-Mars Program for FY 1991. Its chair, Bob Traxler, commented, "This is the camel's nose under the tent. All those space wackos, they want to get started."[110] In a speech at the Marshall Space Flight Center, Bush lambasted the committee for not "investing in America's future" and called them doubters and naysayers. The White House originally planned to pursue a floor fight on the cuts imposed by the House Appropriations Subcommittee, then backed away hoping to make greater inroads in the Senate. But the Senate Commerce Committee approved an authorization bill that cut most of the funding for the Moon-Mars mission.[111] "We're essentially not doing Moon-Mars," Mikulski declared.[112] Existing programs fared reasonably well, but the nose of the Space Exploration Initiative was essentially zeroed out. The following year, when the Bush administration resubmitted its request for funding, only the House Committee on Science and Technology supported the measure. The ancillary policy of the space program continued to place final decisions in the hands of the Congress rather than the president.

Conclusion: Politics Not Science

Pure science and pure politics are two diametrically opposite activities. Decisions in science ultimately imply that there is a proper way to solve a problem. To build a space station, certain laws of physics must be obeyed and certain engineering faults must be avoided. These issues are difficult to compromise. Compromise, however, defines politics. In politics, there is no one proper way to solve a problem. Solutions are defined by counting votes and measuring influence. Politics overrode science in the U.S. space program during Reagan's and Bush's terms, making coherent policy difficult. Scientific questions about shuttle payloads and space station platforms were answered by budget politics, incrementalism, and the decentralized nature of national policy making.

It is unclear whether big science can succeed as ancillary policy, which removes the moorings of long-term planning from policy making and substitutes the politics of the moment. Politics, rather than the substance of the policy, drives decisions. Ancillary policy fosters two opposing sets of political expectations. First, it creates low expectations regarding means. It suggests that only so much will be accomplished at one time and that budget commitments can be adjusted incrementally. Second, ancillary policy creates high expectations regarding ends, for the very reason that the difficulties of achieving them are never fully addressed. Although the government is committed to only one portion of a project at a time, there is nonetheless a vision of what the finished project will look like. But because of the low expectations about means, the vision may be far more grandiose than what can ultimately be attained.

The combination of these antithetical expectations produces a wholly erroneous political maxim that great things (ends) can be accomplished for very little (means). A vicious circle resulted for NASA between what the agency needed to promise to win political support for its programs and what it needed to adequately build and run the programs. Ancillary policy pushed NASA to compromise the feasibility of its projects—in order to obtain the means (political support) to carry them out. Moreover, ancillary policy prohibited presidential leadership by either Reagan or Bush from ending the circle.

In February 1993, looking for ways to cut the federal budget and thereby ease the deficit, a spokesman for the Clinton administration announced that the space station would face the budget axe. The same day, the president assured space advocates that he supported the space station and would not propose its elimination.[113] Clinton's position was an odd compromise between politics and science: there was a continued government commitment to the space program (science), but the commitment was limited by tight budgets and other priorities (politics). Ancillary policy continued to direct the politics of space.

Notes

I would like to thank Roger D. Launius of the NASA History Office for his gracious help in supplying numerous research materials for this project. My sincere appreciation also to Howard McCurdy of American University for his editorial acumen and expansive knowledge of the space program, both of which greatly improved this chapter. My thanks also to Maureen Casamayou for providing an advance copy of her book and to Diana Rix for her research assistance.

1. Reagan, "Remarks at Edwards Air Force Base, California, on Completion

of the Fourth Mission of the Space Shuttle *Columbia,*" July 4, 1982, *Public Papers of the Presidents of the United States: Ronald Reagan, 1982* (Washington, D.C.: Government Printing Office, 1983), p. 892.

2. Senate Committee on Commerce, Science, and Transportation, Subcommittee on Science, Technology, and Space, James Beggs (NASA administrator), testimony, *NASA Authorization for Fiscal Year 1984,* 98th Cong., 1st sess., 1983, p. 51.

3. Reagan, "Remarks at Edwards Air Force Base," p. 870; see also R. Jeffrey Smith, "Squabbling Over the Space Policy," *Science,* July 23, 1982, p. 333; Howard E. McCurdy, *The Space Station Decision: Incremental Politics and Technological Choice* (Baltimore: Johns Hopkins University Press, 1990), pp. 58–62, 139–44.

4. John Kingdon, *Agendas, Alternatives, and Public Policies* (Boston: Little, Brown, 1984), p. 15. Kingdon discusses policies with "high agenda status," which can be contrasted with those with low agenda status.

5. Roger Bilstein, *Orders of Magnitude: A History of NACA and NASA, 1915–1990* (Washington, D.C.: NASA [SP-4406], 1989), p. 58.

6. White House, Office of the White House Press Secretary, "Fact Sheet: U.S. Civil Space Policy," Oct. 11, 1978, p. 1, NASA Historical Reference Collection, NASA History Office, Washington, D.C..

7. McCurdy, *Space Station Decision,* p. viii.

8. George H. Gallup, *Public Opinion, 1935–1971,* vol. 3 (New York: Random House, 1972), p. 2184; Elizabeth H. Hastings and Philip K. Hastings, eds., *Index to International Public Opinion, 1979–1980* (Westport, Conn.: Greenwood Press, 1981), p. 73; Herbert E. Krugman, "Public Attitudes toward the Apollo Space Program, 1965–1975," *Journal of Communication* 27 (Autumn 1977): 87–93.

9. Space Task Group, *The Post-Apollo Space Program: Directions for the Future* (Washington, D.C.: Executive Office of the President, 1969); Beggs, "Why the United States Needs a Space Station," remarks prepared for delivery at the Detroit Economic Club and Detroit Engineering Society, June 23, 1982, NASA Historical Reference Collection, reprinted under the same title in *Vital Speeches* 48 (Aug. 1, 1982): 615–17; NASA, "NASA Establishes Office of Exploration," news release no. 87-87, June 1, 1987, NASA Historical Reference Collection.

10. George M. Low, Team Leader, to Richard Fairbanks, Director, Transition Resources and Development Group, "Report of the NASA Transition Team," Dec. 19, 1980, NASA Historical Reference Collection; National Commission on Space, Thomas O. Paine, chair, *Pioneering the Space Frontier* (New York: Bantam Books, 1986); Sally K. Ride, "Leadership and America's Future in Space," report to the NASA administrator, Aug. 1987, NASA Historical Reference Collection.

11. Advisory Committee (AC), Norman Augustine, chair, *Report of the Advisory Committee on the Future of U.S. Space Programs* (Washington, D.C.: Government Printing Office, 1990), p. 19.

12. Executive Office of the President (EOP), *America's New Beginning: A Program for Economic Recovery* (Washington, D.C.: EOP, Feb. 18, 1981), attachment labeled "Reductions in National Aeronautics and Space Administration Programs,"

pp. 35–36; "Stockman and the 'One-Third' Cutback in the NASA Budget," *Defense Daily,* Jan. 6, 1981, pp. 6–7.

13. McCurdy, *Space Station Decision,* p. 185.

14. Peggy Finarelli to Bart Borrasca, Sept. 8, 1983, NASA Historical Reference Collection.

15. EOP, *Budget of the United States Government: Historical Tables, FY 1996* (Washington, D.C.: Government Printing Office, 1995), pp. 76–81.

16. Bilstein, *Orders of Magnitude,* p. 110, refers to "lean" budgets.

17. For a more detailed analysis of plural institutions, see Lyn Ragsdale, *Presidential Politics* (Boston: Houghton Mifflin, 1993).

18. "Science Panel Losing Thrust in Drive to Shape NASA," *Congressional Quarterly Weekly Report* (CQWR), Sept. 29, 1990, p. 3109.

19. "NASA Cuts Slow Ambitious Plans," *Congressional Quarterly Almanac, 1990* (Washington, D.C.: Congressional Quarterly, 1991), p. 435.

20. In 1984, the House committees passed the authorization bill and the appropriations bill on the same day. In 1991, the authorization committee finished its work before the appropriations subcommittee. In all other years, appropriation decisions preceded the reauthorization bills. See "House Panel Cuts NASA Budget but OKs Space Station Funds," CQWR, Apr. 13, 1991, p. 912.

21. "Tiered-Budget Bill for NASA Gets House Panel's OK," CQWR, Apr. 4, 1992, p. 876.

22. "Science Panel Losing Thrust," p. 3110.

23. "HUD, NASA, Veterans," *Congressional Quarterly Almanac, 1977* (Washington, D.C.: Congressional Quarterly, 1978), p. 284.

24. "Space Station in Trouble Again after Redesign Is Attacked," CQWR, Mar. 23, 1991, p. 743.

25. McCurdy, *Space Station Decision,* p. 130.

26. "NASA Wasted Billions, Federal Audits Disclose," *New York Times,* Apr. 23, 1986, p. A14.

27. Kloman Erasmus, "NASA: The Vision and the Reality," *Report of the National Academy of Public Administration* (Washington, D.C.: NAPA, 1986), p. 42.

28. Joseph J. Trento and Susan B. Trento, *Prescription for Disaster: From the Glory of Apollo to the Betrayal of the Shuttle* (New York: Crown Publishers, 1987), p. 93.

29. NASA, *America's Next Decade in Space: A Report for the Space Task Group* (Washington, D.C.: NASA, 1969), p. 6.

30. Maureen Casamayou, *Bureaucracy in Crisis: Three Mile Island, the Shuttle Challenger, and Risk Assessment* (Boulder, Colo.: Westview Press, 1993), p. 67.

31. John M. Logsdon, "The Space Shuttle Program: A Policy Failure?," *Science,* May 30, 1986, p. 1102.

32. "NASA Cut or Delayed Safety Spending," *New York Times,* Apr. 24, 1986, p. B4.

33. Wernher von Braun, "The Spaceplace that Can Put *You* in Orbit," *Popu-*

lar Science, July 1970, p. 37; Rick Gore, "When the Space Shuttle Finally Flies," *National Geographic,* Mar. 1981, p. 317.

34. Casamayou, *Bureaucracy in Crisis,* p. 68.

35. House Committee on Science and Astronautics, *Authorization Hearings before the Committee on Science and Astronautics for NASA, FY 1974,* 92d Cong., 2d sess., 1972, p. 17.

36. White House, Office of the White House Press Secretary, "Fact Sheet: U.S. Space Policy," July 4, 1982, p. 98, NASA Historical Reference Collection.

37. NASA/DOD, "Memorandum of Understanding on Management and Operation of the Space Transportation System," Jan. 14, 1977, p. 8, NASA Historical Reference Collection.

38. Reagan, "Expendable Launch Vehicles: Presidential Policy Announcement," *Weekly Compilation of Presidential Documents,* May 16, 1983, pp. 721–23.

39. Craig Covault, "Reagan Briefed on Space Station," *Aviation Week and Space Technology,* Aug. 8, 1983, pp. 16–18.

40. This information is found in "Space Program Faces Costly, Clouded Future," CQWR, Apr. 5, 1986, p. 732.

41. Cost figures are all stated in 1971 dollars, much deflated. The real cost of each shuttle flight in the 1992 dollars was set at $393 million. See ibid.; McCurdy, "The Cost of Space Flight," *Space Policy* 10 (Nov. 1994): 277–89.

42. "Space Program Faces Costly, Clouded Future," p. 732.

43. "Pie in the Sky: Big Science Is Ready for Blastoff," CQWR, Apr. 28, 1990, p. 1254.

44. Ibid., p. 1255.

45. Trento and Trento, *Prescription for Disaster,* p. 118.

46. "NASA Cut or Delayed Safety Spending," p. B4.

47. Logsdon, "Space Shuttle Program," p. 1101.

48. "NASA Wasted Billions," p. A14.

49. Klaus P. Heiss and Oskar Morgenstern, "Economic Analysis of the Space Shuttle System: Executive Summary," prepared under contract NASA-2081, Jan. 31, 1972, NASA Historical Reference Collection. See also Presidential Commission (PC), *Report of the Presidential Commission on the Space Shuttle* Challenger *Accident,* 5 vols. (Washington, D.C.: Government Printing Office, 1986), also known as Rogers Commission Report.

50. "NASA Wasted Billions," p. A14.

51. "Space Program Faces Costly, Clouded Future," p. 732.

52. Weinberger, memorandum, Jan. 23, 1984, NASA Historical Reference Collection.

53. "Military Role in Space Program Sparks Concern," CQWR, Nov. 29, 1986, p. 2979.

54. Factors other than the slower than expected launch schedule also inhibited commercial viability. NASA expected microgravity research and manufacturing to lead to commercial applications, but this did not occur, again largely due to cost.

55. Alex Roland, "The Shuttle: Triumph or Turkey?," *Discover,* Nov. 1985, pp. 14–24.

56. "Pie in the Sky," p. 1259.

57. Casamayou, *Bureaucracy in Crisis,* p. 3.

58. PC, *Report . . . on Space Shuttle,* Appendix C, pp. 233–43.

59. Ibid., 1:84.

60. Ibid., 1:245.

61. Ibid., 1:136.

62. House Committee on Science and Technology, *Report of the Committee on Science and Technology: Investigation of the* Challenger *Accident,* 99th Cong., 2d sess., Oct. 29, 1986, Committee print 1016, p. 159.

63. PC, *Report . . . on Space Shuttle,* 1:102.

64. Ibid., 1:82–83.

65. Ibid., 1:84–85.

66. Casamayou, *Bureaucracy in Crisis,* pp. 41–42.

67. PC, *Report . . . on Space Shuttle,* 1:161.

68. Casamayou, *Bureaucracy in Crisis,* p. 42.

69. PC, *Report . . . on Space Shuttle,* 2:F-1.

70. "Space Program Faces Costly, Clouded Future," p. 731.

71. Reagan, "Expendable Launch Vehicles: Presidential Policy Announcement," *Weekly Compilation of Presidential Documents,* May 16, 1983, pp. 721–23.

72. "Veto Message of November 14, 1986," *Congressional Quarterly Almanac, 1986* (Washington, D.C.: Congressional Quarterly, 1987), pp. 35D-36D.

73. McCurdy, *Space Station Decision,* pp. 139–42.

74. "Space Program Faces Costly, Clouded Future," p. 735.

75. "NASA Authorization," *Congressional Quarterly Almanac, 1986,* p. 330.

76. "Science/Technology Authorization," *Congressional Quarterly Almanac, 1987* (Washington, D.C.: Congressional Quarterly, 1988), p. 24.

77. "Space Program Faces Costly, Clouded Future," p. 735.

78. "HUD/Agency Funding Set at $56 Billion," *Congressional Quarterly Almanac, 1986,* p. 172.

79. House Committee on Science, Space, and Technology, Subcommittee on Investigations and Oversight, GAO report, *NASA Program Costs: Space Missions Require Substantially More Funding than Initially Estimated,* 102d Cong., 1st sess., Dec. 1992, pp. 3, 10.

80. AC, *Report . . . on U.S. Space Programs,* p. 32.

81. Ibid., pp. 91–92; "NASA Cut or Delayed Safety Spending," p. B4.

82. Reagan, "State of the Union Message, January 25, 1984," *Public Papers of the Presidents of the United States: Ronald Reagan, 1984* (Washington, D.C.: Government Printing Office, 1985), p. 90.

83. Joseph W. Hamaker, "But What Will It Cost? The Evolution of NASA Cost Estimating," in *Issues in NASA Program and Project Management,* ed. Francis T. Hoban (Washington, D.C.: NASA [SP-6101], 1993); McCurdy, "Cost of Space Flight," p. 282.

84. McCurdy, *Space Station Decision,* p. 171.

85. "Pie in the Sky," p. 1259.

86. McCurdy, *Space Station Decision,* p. 171.

87. "Overhaul Ordered for Space Station Design," *Congressional Quarterly Almanac, 1990,* p. 436.

88. "Pie in the Sky," p. 1258; NASA, "Report to Congress on the Restructured Space Station," Mar. 20, 1991, NASA Historical Reference Collection.

89. "Overhaul Ordered," p. 436; Congress had made a similar, although somewhat less stringent, request of NASA for FY 1988. See "HUD, Agencies Get $57 Billion for Fiscal '88," *Congressional Quarterly Almanac, 1987,* pp. 433–34.

90. Ride, "Leadership and America's Future," pp. 30–35.

91. "Pie in the Sky," p. 1258.

92. McCurdy, *Space Station Decision,* p. 216.

93. "*Freedom* Thwarts Funding Foes; But Bigger Budget War Ahead," CQWR, May 2, 1992, p. 1157.

94. "*Freedom* Fighters Win Again: Senate Keeps Space Station," CQWR, Sept. 12, 1992, p. 2722. Mikulski replaced Proxmire as chair.

95. McCurdy, *Space Station Decision,* pp. 194, 215.

96. Ibid., p. 158.

97. "House Panel Cuts NASA Budget," p. 912.

98. "Space Station Is Questioned as House Passes NASA Bill," CQWR, May 4, 1991, p. 1131.

99. "Panel Protects Domestic Funds by Killing Space Station," CQWR, May 18, 1991, p. 1289.

100. "House Revives Space Station with Cuts to Housing," CQWR, June 8, 1991, p. 1492.

101. "Mikulski Trims Carefully to Pay Cost of *Freedom,*" CQWR, July 13, 1991, p. 1890.

102. Bush, "Remarks at the Texas A&M University Commencement Ceremony," *Weekly Compilation of Presidential Documents,* May 11, 1990, p. 750.

103. White House, Office of the White House Press Secretary, "Fact Sheet: Presidential Directive on National Space Policy," Feb. 11, 1988; McCurdy, "The Decision to Send Humans Back to the Moon and on to Mars," Space Exploration Initiative History Project, Mar. 1992, both in NASA Historical Reference Collection.

104. Craig, interview by McCurdy, Mar. 15, 1991, cited in McCurdy, "Decision to Send Humans," p. 140.

105. National Aeronautics and Space Council, "Final Report to the President on the U.S. Space Program," Jan. 1993, p. iii, NASA Historical Reference Collection.

106. Bush, "Remarks by the President at 20th Anniversary of *Apollo* Moon Landing," July 20, 1989, NASA Historical Reference Collection.

107. EOP, *Budget of the United States Government, FY 1991* (Washington, D.C.: EOP, 1991), pp. 54–56; Bush, "Remarks," p. 749.

108. McCurdy, "Decision to Send Humans," pp. 65–66; National Commission on Space, *Pioneering the Space Frontier,* p. 191.

109. McCurdy, "Decision to Send Humans," pp. 65–66.

110. "Bush Goes on Counterattack against Mars Mission Critics," CQWR, June 23, 1990, p. 1958.

111. "Hubble, Shuttle, and Moon-Mars Add Up to Bad Week for NASA," CQWR, June 30, 1990, p. 2054.

112. "VA/HUD Bill Trims Housing, Boosts NASA," *Congressional Quarterly Almanac, 1990,* p. 860.

113. "Clinton May Not Fight Cuts to Space Station, Atom Smasher," *Arizona Daily Star,* Feb. 26, 1993, p. A4.

6

Presidential Leadership and International Aspects of the Space Program

Robert H. Ferrell

The American space program, an enterprise so wonderful and in many ways so successful, was dominated at the outset by questions of international rivalry and world prestige, and international relations have remained a powerful shaper of the program. There were, of course, more than international reasons to give attention to space—for example, the enormous resources involved. But in the main NASA affected international relations.

And presidents have been key figures in NASA's program. They have tried to guide policy without great involvement by the legislative branch. If an "imperial presidency" has ever existed, it has been in the definition and execution of NASA's international space activities.

Three major projects have been at the front of American foreign policy. At first, there was the intensively competitive Moon race, in which the two superpowers sought to outdo each other. Nothing seemed too much, no step (to employ a primitive, worldly, ground-level measurement) could be missed, to move toward the lunar goal. It was a flag, the American flag, that the astronauts planted on the surface of the Moon when the great moment came in 1969.

During the years of the space shuttle's development, from the 1970s into the mid-1980s, international competition lessened, for after the Americans bested their Soviet competitors in the Moon race the Soviet Union appears not to have participated in any serious race for a shuttle, preferring to develop expendable rocket boosters for space tasks. International cooperation arose between the United States, Canada, and the nations of Western Europe, whose economies were increasingly able to support the high costs of space science and technology.

When it came to constructing and supplying a space station, the principal destination for the American shuttle, the United States chose to emphasize cooperation with its allies—the European Space Agency, Canada, and Japan—in building the large, permanently occupied space station *Freedom*. The Soviets opted for small, occasionally manned, throwaway stations served by expendable spacecraft ferrying occupants between Earth and the station. By the mid-1980s, competition had dropped to a low ebb, ceasing with the collapse of the USSR in 1989. Interestingly, when budgetary as well as technological troubles afflicted the *Freedom* space station, the prospect emerged, mirabile dictu, of cooperating with the Soviet Union's successor organization, the Commonwealth of Independent States (CIS), through use of its latest *Mir* station and, in place of increasingly expensive U.S. shuttle flights, the gigantic rocket named Energiya.

Meanwhile, a technological fallout was coming from the three major projects of the space program with its strong, almost overweening, presidential leadership. Sometimes the successes of a program turn out to be more than the founders envisioned, and such is the case with NASA. In the passage of years into the twenty-first century, the international use of satellites for telephones, television, guidance of ships at sea, weather observation, and managing Earth's natural resources has made a significant difference in the shape of world affairs and in bringing nations together.

International Rivalry

In the beginning was rivalry, the race to the Moon, a project that caught the attention of everyone in the world, not simply the contestants. It appeared to epitomize the very modern world of the years after 1945 when the United States no longer was the dominant power on the globe and was about to receive its comeuppance from the Soviet Union, the nation that defeated Nazi Germany.

Americans now understand that rivalry much better than they once did, for it is now clear that the Soviet Union took its position on space out of weakness. From the beginning, the Soviets were behind in almost every kind of technologically complicated armament. It was the Americans who constructed the first nuclear weapon. When the Soviets exploded their test device in 1949, it was, as we have recently been told, a copy of the Nagasaki bomb, secured through the agency of the German-born, British-naturalized Klaus Fuchs. The Americans managed a huge hydrogen device in 1952 and miniaturized it in 1954. The Soviets did not detonate a thermonuclear device until 1955.[1] Soviet weakness was always there, born of the very nature of the system with its dictatorial ways, which gave Soviet

science an element of rigidity and lack of argument—hence, lack of imagination. This weakness also resided in the economic backwardness of the
Soviet Union, in this respect a developing country, as visitors so often
noticed. When the American H-bomb was tested at Bikini, the United States
was clearly in the lead, and in subsequent weapons races, the results were
always, as Herbert York announced years afterward, a lead time for the
United States over the Soviet Union of between four and six years.[2]

Soviet weakness led to the covering up and a denial of access to most
parts of the country, and on the American side it produced a series of efforts to find out what the Soviets were doing. This led straight into the
Sputnik I launching, proceeded directly from it to the launching of the first
reconnaissance satellite by the United States early in 1961, and in turn
inspired the Soviets to other space endeavors. The moves and countermoves
did not all fit together neatly, but the Soviet accomplishments brought the
Americans into a full-scale, open race for the Moon.

Still, what were the Americans to do? Today, the cold war has passed
into history—the Soviet collapse has inspired more calm appraisals of
events of the last two generations. But for the first fifteen years of the post–
World War II era American intelligence of Soviet missile strength was dan-
gerously poor. Initially, the United States used the Luftwaffe's myriad aerial
maps of the Soviet Union. As a small-fry member of the U.S. Army Air
Forces, I remember a building in northern France full of German maps,
fascinating because of their detail and because they had been printed on
some sort of oilcloth—as tablecloths, they would have prompted exciting
dinner talk. These maps quickly went out of date, and American intelligence officers interrogated the millions of German prisoners who poured
back into West Germany from Soviet areas. Eventually, the stream of re-
turnees dried up, and the next venture was to overfly Soviet borders pho-
tographing obliquely. The Soviets discouraged these enterprises by shooting down a plane now and then.

This was the context for President Dwight D. Eisenhower's administration to welcome the International Geophysical Year (IGY, July 1957–
December 1958) in which both the Americans and the Soviets announced
they would put up satellites. On the Soviet side, the task promised to be
easy, because they already possessed large-thrust missiles. The missiles were
crude, powered by clusters of small engines. On the American side, matters promised to be more difficult, because by this time the United States
had abandoned huge boosters. After miniaturization of the H-bomb and
the Bikini test, such technology was no longer deemed necessary. Thus, any
American satellite would have to be very small, requiring miniaturization

of a complex sort. Eisenhower also emphasized that the American effort would be civilian, not military. The U.S. Army probably had the capacity to send up a satellite as early as the autumn of 1956, but the president refused permission for a military launch.

It is unnecessary to go into the well-known American effort in the mid-1950s to open up the Soviet Union. Walter A. McDougall and others have set out the calculations by which Eisenhower tried to get the Soviets to reveal what he felt but could not be sure was their weakness.[3] The "open skies" proposal at the Geneva Summit Conference in 1955 was part of his strategy. "We knew the Soviets wouldn't accept it," Eisenhower later admitted in an interview, "but we took a look and thought it was a good move."[4] When the Soviets rejected the proposal, the president permitted an extraordinary venture with 516 large weather balloons carrying gondolas containing automatic cameras and radio beacons that allowed tracking, releasing them in Western Europe to float over the Soviet Union, to be captured by U.S. planes when they neared Japan and Alaska.[5] When its results were modest, Eisenhower resorted to the U-2 reconnaissance aircraft.

President Eisenhower had hoped to send over scientific satellites during the IGY, establishing a precedent for military satellites; if the Soviets raised no objection to scientific satellites, military reconnaissance satellites would follow.[6] The Eisenhower administration agonized over how to avoid violating international law and yet get military satellites over the Soviet Union. One suggestion was to send them first into orbit over the friendly political skies above the equator, then send them north. The Soviets solved this dilemma by lofting *Sputnik I,* ignoring the legal tangle and justifying their action by asserting the right of vertical freedom of space and, as for the horizontal, claiming that their satellites did not fly over countries below but the countries themselves rotated under *Sputnik.*[7]

If the *Sputnik* crisis resolved the Eisenhower administration's legal problem, it nonetheless produced a cornucopia of cold war confusions. At the outset was U.S. consternation that "the Russians are ahead!" People counted engineers, of which the Soviet Union possessed hundreds of thousands. American college and university administrators jumped on the *Sputnik* bandwagon, securing the National Defense Education Act (NDEA), a thinly disguised piece of legislation that brought all sorts of studies under federal subscription. I remember my personal confusion one day to find that a close friend, a folklorist, had obtained a series of fellowships for his graduate students under the NDEA. On the national scene, Eisenhower sought to quiet the uproar, only to find his words lost in the public melee. The Democratic party mercilessly berated the nation's oldest (and previously quite popular) president, unconcerned that Eisenhower's health was parlous—a heart attack in 1955 (and probably two earlier ones in 1949 and 1953), Crohn's disease in 1956, and a stroke in 1957. The most opportunistic American political leader in many years, Senator Lyndon B. Johnson, took the advice of George Reedy and began to use *Sputnik* to gain the White House, but he was forestalled for a short time by another Democratic politician who managed to speak more convincingly of a "missile gap" and, like Johnson, made little effort to find evidence to the contrary.[8]

Meanwhile, Nikita S. Khrushchev was playing one of the more remarkable international shell games in the history of the twentieth century. As in the case of Communist China during this era (critics said that China was weak but Mao was strong), so with the Soviet Union and Khrushchev, although in no sense either so obviously or so completely. His assertions constantly kept the Americans off guard. During the Suez Crisis of 1956, the Soviet Union threatened the Western allies through graduated messages to their leaders, scaring the weak French government by referring to a possible deluge of intermediate-range rockets, telling the wavering British government that only a few warheads could remove the British Isles, and informing the American government with studied innocence that both the United States and Soviet Union possessed nuclear warheads and the means to deliver them. In one of his Kremlin tirades, Khrushchev avowed that his country possessed missiles of such accuracy that they could hit a fly in outer space. With these pleasantries raining down upon the governments of Western Europe and especially upon Washington from 1956 until his disappearance from the international and national scene in 1964 (under the claim of "adventurism" by his successors), the Soviet premier did not hesitate to accompany his bluffs by producing crises in Berlin and Cuba, not to mention stirring up developing countries everywhere.

This, then, was the milieu in which, after the special challenge of *Sputnik* and of *Lunik* in 1959 (sending an unmanned spacecraft straight into the Moon), came the flight of Yuri Gagarin in April 1961, shortly after President John F. Kennedy's inauguration and shortly before the Bay of Pigs affair, twin humiliations that almost certainly fostered the Moon race. The manner in which Americans accepted Khrushchev's exaggerations—for they failed to sense that if the Soviet Union was ahead in some space exploration the space race as a whole had hardly been decided—was extraordinary, although of a piece with such previous American effervescences as the fear of Native Americans, the Palmer raids after World War I, and

McCarthyism. This became evident when, nearly a year after Gagarin's flight, John Glenn spent five hours in space on February 20, 1962. Indeed, the resultant enthusiasm of his fellow citizens from Ohio sent him to the Senate, where he still remains.

The Race to the Moon

Many books and articles have described the Moon race in all its particulars and massive achievement.[9] Even now, the photographs taken by the spacecraft, as men and machines came ever closer, not to mention the first landing and its subsequent five missions to the surface, are of absorbing interest. The literature describes how the grand enterprise began in 1958 when manned landings began to be discussed during the transition from NASA's predecessor organization, the National Advisory Committee for Aeronautics (NACA), to the organization of NASA, and then the giant step forward with President Kennedy's formal decision to go to the Moon in 1961. The race to beat the Soviets to the Moon encompassed eleven years. It consumed the attention of everyone in NASA, which at its peak employed 36,000 people in civil service jobs and contractors hired 400,000 more workers. During the race, the agency delayed all other programs, and general space sciences garnered only a modest budget.[10]

At the outset, the cost for a manned Moon landing was anyone's guess. Critics were claiming $100 billion. NASA administrators made no effort to guess low, and the last director of NACA, who was NASA's first deputy administrator, Hugh L. Dryden, estimated a cost as high as $40 billion. Shortly after Gagarin's flight, Robert C. Seamans Jr. estimated the cost at $20–40 billion. NASA eventually settled on Seamans's figure and to hold to it demanded flexibility and adherence to its own timetable. In congressional hearings, NASA administrators pointed out that each year of delay would cost $1 billion. In the end, the project suffered a three-year delay, half of it because of the 1967 accident in which three astronauts lost their lives on the launchpad; the delay made a project total (up to the time of the first Moon landing but excluding the subsequent five landings) of $23 billion.[11]

During Project Apollo there was never any question of cooperation with the Soviets, save occasional talk including a curious speech by President Kennedy before the United Nations in September 1963, in which the president remarked to no one in particular, "Let us do the big things together." His comment elicited no response from the other side other than dismissals by a few Soviet editors who pronounced it "premature."

For a variety of reasons, Americans did not attempt to bring the Western Europeans or Japanese into the project. NASA's first administrator,
T. Keith Glennan, told the hardheaded director of NASA's international
relations, Arnold W. Frutkin, that "international cooperation might, in the
end, make more sense than any other aspect of the program." As Frutkin
remembered, Glennan "said it in just so many words."[12] In truth, there
was not much Frutkin could do with an international program beyond
arranging for tracking stations and otherwise bringing in foreign technicians and students for American university training. Apollo had to be an
American show; the technology abroad was not advanced enough to use,
and the international rivalry at its heart required a U.S. demonstration
project. In the only two European countries with serious space projects,
France and Britain, special conditions obtained. The French government
under President Charles de Gaulle was trying to exert its own and Europe's
independence of the United States, and Washington could not cooperate
with Paris without proliferation and competition, with some of the so-
called "dirty interfacing" (revelations of American technology) going off
in the direction of the Soviet Union. As for the British, when *Sputnik* went
up they threw in the towel and canceled their intermediate-range missile
project, the Blue Streak, hoping that the United States would give them a
similar missile, which they needed to extend the life of their aging V-bomb-
ers. When the Kennedy administration's Department of Defense (DOD)
under Secretary Robert S. McNamara canceled the substitute missile program, Prime Minister Harold Macmillan arranged with his American counterpart to obtain Polaris missiles for Britain's nuclear submarines. This
diplomacy diverted any British interest in Project Apollo.

For such reasons the Moon race was an American-Soviet affair. In a
notable debate in the U.S. House of Representatives shortly before the
triumph of Apollo, the point of view became apparent. Some of the ora-
tory was due to congressional tradition, wherein an abundance of words
sounds good to the folks back home. Some of it derived from the excesses
of the speakers. In this regard, Richard L. Roudebush (R-IN) set the tone.
He offered an amendment to NASA's authorization bill in 1969, requiring the implanting on the surface of the Moon of the American flag and
none other. A similar amendment had disappeared in committee because
it implied that the United States was about to establish sovereignty, that
is, ownership of the Moon. That would have violated the Outer Space
Treaty of 1967. Roudebush altered his own amendment to show that
implanting the flag was "intended as a symbolic gesture of national pride
in achievement and is not to be construed as a declaration of national

appropriation by claim of sovereignty." The legislator spoke with Hoosier eloquence:

> Over $23 billion in hard-earned taxpayers' money will have been spent to carry out this formidable task. In all due fairness to the American taxpayer, it does not seem too much to ask that our flag—Old Glory—be left on the lunar surface as a symbol of U.S. preeminence in space to which the citizens of this Nation can refer with pride. . . . History and national pride dictate that our achievements be duly commemorated. I know of no act more significant nor symbolic that would memorialize our achievements than the erection of the "Stars and Stripes" on the surface of the Moon.

The Soviets, Congressman James G. Fulton (R-PA) pointed out, "recently sent the coat of arms as well as a picture of Lenin to the surface of Venus." Congressman James Symington (D-MO) countered that through NASA the president should direct what flags or symbols should be placed on the Moon; depositing Soviet symbols on Venus was a bad idea because, he noted,

> I do not recall that this occasioned the general approbation of mankind. Nor did I realize we were accepting lessons from that particular source in how to win the hearts and minds of men. . . . Jefferson wanted us to maintain "a decent respect for the opinions of mankind." What "respect" does this graceless edict demonstrate for the opinions of nations which produced Galileo, Copernicus, Newton, Einstein, Tsiolkovsky, and other giants in thought and deed? What star or stripe is tarnished on Old Glory by a simple gesture honoring the whole history of man, his collective dream, and his epic persistence without which our own continent might yet be undiscovered?

Other congressmen pointed out that the flag was already on the Moon by being painted on the *Surveyor* spacecraft. Hence another flag would not hurt anything. Allard Lowenstein (D-NY) thought the president should make the patriotic choice.

To all this point-making Roudebush was oblivious. "I feel compelled," he announced, "to offer this amendment in view of the many proposals being put forth which advocate that our spacecraft carry to the surface of the Moon the United Nations flag, the flags of other nations, or other emblems or articles symbolic of international cooperation in space exploration." The Roudebush amendment carried on a voice vote, with an overwhelming chorus of "ayes."[13]

Some years later, perhaps because of the troublesome end of the Vietnam War, with patriotism worn a bit thin, and as a contribution to détente, the United States and the Soviet Union agreed to conduct a joint spaceflight, known as the Apollo-Soyuz Test Project (ASTP), in which a three-man *Apollo* crew docked with a two-man *Soyuz* crew. The USSR agreed to a few modifications of the *Soyuz* to permit docking and conducted a test flight in 1974. Both countries launched spacecraft on July 15, 1975. The two ships docked on July 17, exchanged visits and joint experiments, and undocked on July 19, with *Soyuz* returning to Earth two days later, *Apollo* three days thereafter. Because the experiment marked the last flight of the Apollo spacecraft, the androgenous docking adaptor became obsolete at the end of the mission.[14]

In the short term, Project Apollo was an American triumph. In the long term, the cost, $23 billion, was large and might have made a difference in President Johnson's Great Society programs or helped avoid the inflation that fueled dissatisfaction with the Vietnam War.

On the American side—and another minus—it is possible to contend that the Moon race led straight into defeat in Vietnam. When the U.S.-Soviet rivalry turned technological, it combined with other factors to confuse the U.S. officer corps, making them think that technology would win any sort of war, such as the conflict in Vietnam. The Kennedy administration, to be sure, talked a great deal about winning the hearts and minds of the people, of counterinsurgency. But the generals and colonels often thought in terms of winning with technology.

Indeed, the large admixture of bluff in Soviet policy under Khrushchev may have delayed the coming of real détente by ten or twenty years. Khrushchev's bluff with *Sputnik* was caught by the first American spy satellite in 1961. The United States announced the fact, throwing the truth in Khrushchev's face. There followed the Cuban missile crisis and Khrushchev's removal from power in 1964 and his replacement by hard-line successors.

Interestingly, the ultimate American technological confrontation with the Soviets, the Strategic Defense Initiative, was not meant to be a bluff. President Ronald Reagan was a believer. Yet the scientific impossibility of the program made it a bluff. In that sense, the Americans borrowed, with enormous success, from the lexicon of the Soviets.

The Space Shuttle

After Project Apollo, the great American space triumph in the cold war, NASA turned to the second of what proved to be three large projects, construction of a space shuttle, sometimes known as the Space Transportation System (STS). Given that Apollo was a project for humans in space, not machines, it was only to be expected that the shuttle, sponsored by President Richard M. Nixon in 1972, looked in the same direction. As happened with Apollo, so with the shuttle: scientists by and large did not favor it, believing that instrumented packages would perform just as well, would be cheaper, and would not run the mortal risk of failure. But the "man in the Moon" dream captured Americans and Soviets alike and Apollo continued to hold public attention, with the result that the space choices seemed to be the shuttle and after it a space station. More distant space shots and a "grand tour" of the planets was a distinctly third choice. With money tight because of the Vietnam War and the Great Society programs, NASA began with the shuttle.[15]

NASA's decision to develop the shuttle defined what was to be the principal international component of the project, what became known as *Spacelab*. From the beginning, *Spacelab* was an international project. Its origin sprang from an understandable NASA effort to involve Western Europeans in a project that would have a "clean interface," that is, it would not reveal American technology that might be passed to the Soviets. This meant a module or, in a sense, one "can" that would fit within another "can," the shuttle.[16]

Spacelab, one should add, was (to mix the figure) a bit of a smoke screen, and NASA was little fascinated by it. But the agency needed *Spacelab* as a purpose for the shuttle, apart from transport of satellites, for when shuttle costs inevitably began to escalate it would be clear that rockets could carry satellites much more cheaply then the vaunted STS. Thus, NASA needed *Spacelab* until Congress appropriated the money for the space station, after which the shuttles would become freighters for the station.

Persuading the Europeans to create a justification for the shuttle by constructing a "research and applications module" was fairly easy, for they too needed to gain time for a space project of their own. With the approaching quincentennial of Christopher Columbus's first voyage to America, it evidently occurred to Europeans that the shuttle module could be the precursor of a several-module European space station that could bear Columbus's historic name and restore Europe to the primacy it had lost centuries before after a tricky Genoese did some fast talking with two simple-minded Spanish monarchs. In any event, the twentieth-century Columbus scheme never came to pass, because to pay for it—including a proposed European shuttle known as *Hermes* (why not Amerigo Vespucci?)—would have cost $6 billion for *Columbus* plus $9 billion for *Hermes.* The abortive project would truly have required Queen Isabella's crown jewels.

For a while, the Americans had to deal with the predecessor organizations of what in 1975 became the European Space Agency (ESA): the European Scientific Research Organization (ESRO), for spacecraft, and the European Launcher Development Organization (ELDO), for launchers. Unfortunately, these organizations were of questionable competence. ELDO was trying to build a three-stage booster, each developed in a different country, a sure-fire prescription for failure. Coupled with this disorganization was the European way of doing things. On one occasion, an American member of a team dealing with Italian scientists at Turin proposed a working lunch of sandwiches. As he later wrote, the Aeritalia team leader, Professor Vallerani, reacted first with shock, then dismay, and finally disbelief. The professor noted several times that "it *could* be done, it would just take *one* phone call, was that *really* what we wanted to do?" Finally, the professor exploded, "Yes, it can be done—but it's never *been* done!" The visitors accepted plans for a modest lunch of five courses.[17]

Despite the awkwardness of cooperation, it seemed worth the effort. Hans Mark, a NASA deputy administrator in the early 1980s, flew to Europe with a Boeing 747 carrying the test shuttle *Enterprise* and discovered the excitement of Europeans over what the American space program was about to do. The 747 pilot overflew London and received permission from the Heathrow tower to fly down the Thames River at three thousand feet. The result, as Mark described it, was "absolutely fascinating. It was Sunday afternoon and many thousands of people were lining the river watching for the *Enterprise* to pass overhead. The crowds were enthusiastic—even at that altitude we could see them cheering and waving. Then, in the course of three minutes, we flew over the Parliament at Westminster, the Tower, and the famous observatory at Greenwich." Nor was that the end, for the landing at Stansted was "completely overwhelming." People held their hands up making the "V" for victory sign, and many were in tears. "I cannot explain why this happened, I can only record it." The *Enterprise* was the star attraction at the Paris Air Show. On return it stopped in Ottawa, the Canadian capital, where 400,000 people turned out to see it.[18]

Several years later, the Europeans discovered the huge costs of venturing into space. The Americans could remember the $23 billion cost of Project Apollo and now found that just to enter the shuttle test phase cost $5.15 billion, with only two shuttles purchased out of a proposed fleet of five, not to mention launch costs. The Europeans were shocked by the unending requirements of *Spacelab* for documentation, interfacing, and testing. By 1982, the cost of *Spacelab* had grown from $250 million to $1 billion.

Then came two more shocks for the Europeans. First, in a budget bind, NASA canceled a U.S. spacecraft that was part of a two-spacecraft International Solar Polar Mission to observe both poles of the sun. The agency had enlisted the Europeans in the project but canceled with little more notice than a telephone call.[19] Then, after airily giving ESA officials the notion that it would assume the cost of *Spacelab* by buying a half dozen copies, NASA bought one unit at a cost of $128 million.

As if these experiences in international cooperation were not enough, flight charges on the shuttle turned out to be far higher than expected, advancing from $10 million to $300 million, so that ESA could not afford to send up its module. The United States paid for the first flight, the Germans the others. The Europeans complained that they could not even afford to pay for in-flight experiments. The cost of shuttle flights created an impossible situation, which NASA officials could not easily explain. ESA accused NASA of including in its charges all its bureaucratic overhead at Kennedy Space Center (which may have been true). "We tried to explain to them that we had developed the Space Shuttle, and yet users in the U.S. government paid the same price that was charged to ESA," wrote Douglas R. Lord. "Somehow this argument was never accepted by our European friends."[20]

Concerning international participation in the shuttle, suffice to say that only one project proved satisfactory and that was Canadarm, the $100–million remote-controlled crane carried in the shuttle and used to manage satellites. It amounted to an arrangement with Canada whereby America's northern neighbor obtained all the tickets for shuttle flights it desired. The Canadians were happy, and so was NASA, which received a free crane, and the interfaces were as clean as those for *Spacelab*.[21]

For the rest of the involved nations, the space shuttle had little to do with international relations other than carrying international passengers, which scientifically was no more of an achievement than, say, the Soviet launching of *Sputnik* in 1957 and of much less international interest. In 1983, with the first *Spacelab* mission, the shuttle took up a West German astronaut, for West Germany had footed most of the cost of *Spacelab*. He was followed by Canadians, Europeans, a Japanese, and a Saudi prince.

Space Station *Freedom*

The next major project of the American space program, the space station, was to have more of an international component that *Spacelab* and considerably more financial involvement. Because of a series of confusions over planning for the space station, its development raised more questions about cooperation. After the breakup of the Soviet Union, it presented fascinating possibilities for CIS-U.S. collaboration.

When a new NASA management team took over in 1981, Administrator James M. Beggs and his deputy, Hans Mark, announced at their confirmation hearings that their top priority was the space station. From that moment onward, it was presumed that the station would possess a large international component. The reasoning was necessarily complicated. For one thing, there was much international idealism among the American people; after the public indifference that had plagued development of the shuttle, NASA needed all the support it could muster. In particular, there was the ever-present money problem. From the beginning of the shuttle project, the space station had been on hold because of limited funding. It was impossible to propose the station and the shuttle at the same time, because there was not enough money, so the shuttle came first. That made the shuttle embarrassing, for it needed a place to go. *Spacelab* was not enough of a place. The shuttle could not be a boxcar for satellites, for a rocket could do that job much more cheaply. As soon as possible, NASA administrators had to advocate the space station, to get the boxcar problem out of sight. But then the money problem reappeared. In 1981, no one spoke of the possible cost of the space station because no one knew. That was where the international side of the station came in. If the cost was large, it would be better to spread the cost. Also, what better way to forestall competitive European and Asian programs than to co-opt them with the American space program?

There was yet another reason for bringing in other countries. As Captain Robert F. Freitag, an early space station advocate who could be counted on to describe the issue with style, recalled, "We knew that if we found ourselves locking in with international agreements, it would be awfully hard to say no to the program."[22]

To give them credit, NASA administrators were not entirely cynical about international participation. During the 1970s, they had seen how the primitive technological accomplishments of Western Europe and Japan had given way to prowess that was close to the abilities of the world's premier technological power. Whatever the concern about interfaces that had marked the shuttle program, NASA administrators knew that sooner or later the Europeans and Japanese would catch up. The world was equalizing in technological ability, whatever nations did to prevent it. They might as well embrace international cooperation because they would be forced into it anyway. And, lastly, they were all products of the 1940s and thereafter and knew that the old isolationist days of the 1920s and 1930s were gone. Adolf Hitler's government in Germany had been defeated only by an Allied coalition.

Cooperation, once undertaken, had its dangers, for within the United States the 1980s were the era of Reaganite conservatism, and the administrators had to guard against any claim of "dirty interfaces." A group known as the Space Station Technology Steering Committee scheduled a meeting at Williamsburg, Virginia, in March 1983, for a discussion of "recommended advanced technologies." A pro forma invitation was issued to foreign participants. But NASA administrators withdrew the invitation, worrying that it might lead to "a massive hemorrhage of U.S. technology."[23] By mid-1983 it became evident that the potential of the station for international involvement would not sell it in Congress, and Beggs and such lieutenants as Kenneth S. Pedersen and Margaret Finarelli retreated on international cooperation, not exactly trying to get away from it but not stressing it as a selling point. To Congress they offered the space station, eventually christened *Freedom*, as a modern-day Fourth of July.[24]

Another awkwardness was the sudden hostility of the American military establishment, in the person of Defense Secretary Caspar Weinberger, who insisted that his department wanted nothing to do with *Freedom*, a statement that was deeply embarrassing because the military was expected to pay for many shuttle flights. The good side of this disappointment was that NASA, which had been courting the military, could now avoid it and please the West Europeans and Japanese who would find a strictly civil program more appealing. NASA decided, in Beggs's words, to ask the president to approve "a completely civil station." ("Keep in the idea of international participation," Beggs added, albeit to members of his space station task force.)[25]

It was necessary to pooh-pooh the refrain of the large body of scientists who said that everything a space station could do a shuttle with a lab module could do better. *Freedom*'s chief scientist, Robert W. Phillips, denied this; he admittedly said so in 1992 but nonetheless echoed the explanations of a decade before. "The plant people can't wait to do a seed-to-seed experiment," he said. "You can't do that in a week." Same thing for animals: "Every organism I know has been changed in space. Not to mention the advantage of forming materials under conditions where separation in a mixture is no longer based on density."[26]

Lastly, in making the space station attractive NASA authorities estimated its cost at an absurdly low figure of $8 billion. The reason, the director of NASA's space station task force, John Hodge, recalled, was that "I reached the scream level at about $9 billion."[27] They omitted a few small things, such as the cost of transporting the station from Earth to orbit, operating it once in orbit, and conducting the experiments with seeds, rats, and crystals. They put in an impossibly small reserve for changes and limited spending for ground support. They said nothing about a "lifeboat" to get the crew back to Earth if the station became something less than a station. With these things tucked away, Beggs ingeniously announced that the president could purchase space station *Freedom* "by the yard," buying and sending up a piece, presumably, when the mood inspired him.

This set the stage for President Reagan's announcement of the station. He decided to support the station early in December 1983, although the decision as to its international proportions seems not to have been his, except in a formal sense. In a meeting at the White House on January 18, 1984, without the president, an ad hoc group that included Beggs decided to place the space station and international participation in it in the State of the Union address scheduled one week later. The president's "speechwriting office," according to John M. Logsdon, got in touch with NASA to fit in the appropriate words. On January 25, the president informed Congress and the nation, "Tonight, I am directing NASA to develop a permanently manned space station and to do it within the decade." Later in the speech he included the international words: "We want our friends to help us meet these challenges and share in their benefits. NASA will invite other countries to participate so we can strengthen peace, build prosperity, and expand freedom for all who share our goals."[28]

The White House group decided that Beggs would carry the presidential invitation to foreign governments, and the administrator did so, like a global traveling salesman, aboard a White House plane. An economic summit meeting in London was planned for June 1984, and a group of subofficials known as the "summit Sherpas" arranged for their principals to declare at the meeting that they "agree in principle to cooperate in the development of an international Space Station, demonstrating that free nations will continue to use outer space for peaceful purposes and for the benefit of mankind."[29] The summit leaders met and, with the key words spoken, strolled out of the meeting room to encounter, on a table, a model of the space station. Photographers captured the moment, with Reagan standing in front of the model, arm extended, explaining the station, Prime Minister Margaret Thatcher standing next to him looking closely at the model, and Prime Minister Yasuhiro Nakasone on the other side of the table looking on, lips pursed.

In working out international cooperation on the station over the following years, the enterprise was made much easier by the piecemeal approach. Whatever the eventual appearance of space station *Freedom*, it was like the children's game of a generation or two ago: a celestial Tinkertoy, its modules, solar panels, and other components fitting together any way the United States and cooperating nations decided. Because of experience with *Spacelab*, the West Europeans in ESA chose to contribute a service module. Because of experience with the shuttle's Canadarm, the Canadians proposed a bigger arm, a Mobile Service System.

In the late 1980s, the several projects seemed to be going along well, with a few problems, as one might have expected. But the funding issue continued to haunt the effort. The cost of the European module rose to $4.5 billion and the Japanese to $2 billion. In the European case, the opportunity to borrow equipment in *Spacelab* did not seem to make much difference. In the Japanese case, it may have been the novelty of everything. For the Canadian arm, the Ottawa government expected to spend $1 billion, ten times the cost of Canadarm. It was to be a much larger arm, 58-feet long, with a payload of 128 tons. Two smaller arms would contain up to nineteen joints to perform finely detailed tasks that otherwise would have to be done during a spacewalk by an astronaut. The price might increase even more because of the awkwardness of using the arm in the extreme darkness, or brightness, of space, which skewed human estimates of speeds and distances (experiments by shuttle crews had illustrated this). The arm might require a robotic vision system with a computerized eye. The size of the space station might require more repairs than had been estimated, and for this the station crews would need the arm equipped with all possible gadgetry.[30]

Incidentally, discussion of the Canadian arm became awkward for NASA officials, for in advancing the several reasons for the arm—that spacewalks by astronauts were dangerous because of radiation, the risk of being stranded, and the possibility of being struck by floating debris— they brought into public view the debris issue, which was serious. The proposed station would be a large structure, and it might be struck by floating debris. Indeed, here lay the need for a particular sort of international cooperation in rules to halt the dumping of debris in space. One piece of debris the size of an aspirin tablet travels at 22,000 miles per hour and packs the punch of a 400-pound safe moving at 60 miles per hour. More than 3.5 million man-made objects are whirling in orbit and could force the proposed station to use shields. The General Accounting Office (GAO) estimated a 36 percent chance that a piece of debris would strike a critical part of the station and that redesign to provide shields could lower the risk to 12 percent. The GAO report called on NASA to halt design work until it addressed the shield problem.

Over the entire plan for international cooperation lay the complexities of dealing with four groups of space station participants—Americans, Europeans, Japanese, and Canadians. Cooperation virtually required a manual, a variant of *Roberts' Rules of Order,* before any kind of conversation could begin. When an issue arose, there was extreme awkwardness in how to make decisions. There were three program coordination committees: U.S.-ESA, U.S.-Japan, and U.S.-Canada. That is, at the outset, all three groups had to approve a joint program plan. Mundane decisions might be fairly easy, but for larger decisions there were so-called memoranda of understanding (MOUs). Here was more confusion because of a double standard of promises. On the American side, an MOU meant only an agreement with NASA. For the international partners, it meant an intergovernmental agreement (IGA) that carried the force of a treaty. This put the United States in the awkward position (for its international partners) of exchanging simple promises for ironclad, binding agreements. If NASA decided to go back on an MOU it could do so with a bow and scrape and no more; the other nations would have to renegotiate everything with their presidents, premiers, and parliaments.[31]

This addressed nothing of possible conflicts among international crew members operating a future space station. On this subject, the possibilities could not be overlooked, according to the psychiatrist Patricia Santy of NASA's medical sciences division at the Johnson Space Center in Houston. According to a 1988 IGA, the United States would have criminal jurisdiction aboard the space station *Freedom.* But the American government was to consult with the miscreant's government before proceeding to trial. A designated space station commander, a career astronaut, not a scientist, would have final authority in resolving operational disputes. Hence, crew members needed to know about foreign cultures, related Santy: How long are you in the shower? How and what do you eat? Because of the heterogeneity of residents aboard space station *Freedom,* NASA in 1991 drafted a code of conduct.[32]

A question arose about press relations aboard the future space station. Cable News Network (CNN) approached the administration of President George Bush about the possibility of a small bureau on space station *Freedom.* What if Reuters asked for a bureau?

Congress and the Space Station

As if all these complications did not suffice, Congress in 1990 informed NASA of the need to cut $6 billion from the space station project over the next five years. This was serious. The cost of the project had risen from $8 billion to $14.5 billion and then soared to $38 billion (including the international contribution). There were two ways to make the cuts required by Congress, and NASA chose both. The first was to stretch out the program, with "milestones" or points of achievement farther apart. This meant milestones well beyond the turn of the century, far beyond President Reagan's goal of 1994. The second was to reduce the size of the station, which NASA redefined from a projected 508 to 353 square feet. Reduction in size meant fewer shuttle flights to take up Tinkertoy components of the station, six instead of eight electricity-producing solar panels, and perhaps fewer experiments. It meant fewer people in the crew and less water to transport on the shuttle (the water requirement every three months would now be only nine tons). A smaller station also meant less maintenance—fewer adjustments with the Canadian arm.

Reducing the size of the station—known felicitously as rephasing, descoping, or restructuring—produced mixed feelings among the space partners. A distinct point of irritation was that NASA announced the smaller station without consulting its partners, an act reminiscent of the International Solar Polar Mission a decade before. Perhaps this was because the partners were bound by IGAs and NASA by MOUs. For the ESA partners, downsizing was nonetheless satisfactory. They were having budget woes, especially the West Germans who were trying to swallow East Germany. "If NASA is going to reduce their space station, I think we have the right to reduce ours as well," said one German engineer, forthrightly, speaking of the ESA module, mostly German financed. For the Japanese, downsizing brought approval, if for a different reason. When a Japanese Space Agency official, T. Kato, said his team was "moving pretty good," there was a touch of national pride in his voice. "We didn't change any lengths. Ours was to be the shortest module. Now it will be the longest."[33] For Japan, however, stretching out the milestones was worrisome, because delays would increase costs.

Then, in February 1993, came another order to rephase, descope, and restructure, this time not from Congress but from the incoming administration of President William J. Clinton. NASA officials took up the task, with a promptness that was almost embarrassing, making observers won-

der why NASA needed even the 353-foot station. Administrator Daniel
S. Goldin announced, "We stand at the doorstep of an incredible opportunity." Under the first downsizing, the space station would have taken
more than a decade to build. "If we continued with huge long-term
projects," he said, "the technology will always be outdated. Do we want
to be up there with something in the year 2030 that was designed in the
1980s?"[34]

But the 1990 congressional reduction—what one wag described as the
equivalent of NASA's moving from the Hyatt Regency to Motel 6—and
the Clinton administration's 1993 downsizing came just at the time when
the Soviet Union fell apart, and an altogether unexpected development
forced itself into the calculations for space station *Freedom:* the Russians
promised to come to NASA's assistance in solving some, perhaps all, of
the station's problems. The easements the CIS nations proposed displayed
some of the Soviet space successes and gained support from America's
space allies who, if the truth were told, had never been enthusiastic for
the space station.

The CIS nations promised to get NASA out of all its predicaments. The
space station could be a complete turnaround from the Moon race of the
1960s. First, they presented a way out of the nearly impossible cost of
shuttle launchings. Downsizing the space station meant fewer launches—
all to the good, but not enough to count. Simple maintenance of facilities
at Kennedy Space Center so raised launch costs that, when *Endeavour* in
1992 fixed a rogue satellite that itself cost $270 million to build and launch,
it would have been cheaper to throw the satellite away and launch a new
one.[35] The DOD paid for ten shuttle missions, contrary to Weinberger's
injunction, but in 1992 DOD-paid missions ended. The possible CIS solution to shuttle costs was a twenty-story rocket, Energiya, which first flew
in 1987. One Energiya might cost $500 million, and it was estimated that
four, accompanied by four shuttles to unload the freighters and assemble
the space station (cost of each shuttle flight: $1 billion) would do the job.[36]
Total cost would be $6 billion, compared to twenty-two shuttle flights ($22
billion). There were two other pluses. Energiya would avoid the problem
of space station components being "shuttle compatible," that is, fitting into
a cylinder the size of the shuttle bay. It would avoid the distinct possibility of another *Challenger* accident if the shuttle flew twenty-two flights to
loft the station and ten more to supply the station once operational. But
there also were minuses. NASA engineers had spent years making the space
station shuttle-compatible and to use a heavy-lift vehicle would force redesign of the station in order to assemble and integrate larger pieces on

the ground. Moreover, only two Energiyas had been tested. Energiya uses four one-stage Zenit rockets as its own first stage, and in separate launches Zenits have failed several times. Zenit also is produced in the Ukraine, and if any trouble occurred between Ukraine and Russia (developer of Energiya), the Energiya Program would have to terminate. Moreover, utilizing Energiya meant taking station components to Tyuratam in Kazakhistan and using Russian ground-launch crews—would Russian political stability last for the life of the space station, an estimated thirty years?[37]

A second Russian suggestion to NASA was the possibility of junking space station *Freedom* and using the Russian *Mir* station. The cost of the American space station was driving the Europeans in that direction, because of the danger (like *Spacelab*) of no money for experiments in the *Columbus* module (ESA had retained the name of Columbus from the once-talked-about ESA space station) of space station *Freedom*.

Why not take *Mir*? Russian stations, of course, were small, and not built to last thirty years. They were only occasionally, not permanently, manned. But NASA, in opting a second time for a smaller station, this time under President Clinton's direction, said that the smaller station would not have a permanent crew anyway. Innumerable people had lived in the *Mir* stations, including a representative of almost every nation in the erstwhile Eastern European bloc. The Russians held the space endurance record by a wide margin: 366 days. The ESA announced plans to use *Mir* to train and launch astronauts. In 1990, the Japanese Broadcasting System (not the Japanese government) reportedly paid $12 million to send a journalist, Toyohiro Akiyama, on a trip to *Mir*. The logical next step would be to propose substituting *Mir* for *Freedom*.[38]

Another Russian solution to a NASA problem was the Assured Crew Return Vehicle, or lifeboat. When NASA officials proposed the space station, they said nothing about the lifeboat. Nor did they say anything about it when they scaled down the station in 1990, for the scaledown reduced crew size, in between shuttle visits, to four. They could have set the figure at four on purpose: the Soviet lifeboat, the *Soyuz TM*, takes three. NASA had studied a four-person rescue vehicle but estimated its cost at $1.6 billion. The *Soyuz TM* would cost $30 million, not counting adaptation costs. Unfortunately, to land the *Soyuz TM* requires a land base of at least nineteen square miles, within certain latitudes. There are no acceptable sites in the United States, where the only areas in range are southern Texas and Florida. But Australia would be possible. After NASA in 1992 signed a $1 million contract with NPO Energiya, builder of the Soyuz rescue vehicle, forty engineers and managers met in Houston to discuss the lifeboat issue. Russian participants in working group sessions were previously involved with the Apollo-Soyuz Test Project in 1970–75. It was their first visit to the Johnson Space Center since that time.[39]

International Commercial Applications

Begun in U.S.-Soviet competition, the space age showed possibilities far beyond cold war issues. The commercial advantages of space technology attracted international attention in the early 1960s, even before the U.S. triumph with the Moon landing.[40] In 1962, NASA launched the first two international satellites: *Ariel,* with a scientific payload for the British, and *Alouette,* for the Canadians. That same year President Kennedy signed the Communications Satellite Act, which established the Communications Satellite Corporation (COMSAT), to cooperate with other countries in producing an operational system and to provide services to other countries. Then in 1963 came the beginning of an almost revolutionary use of communications satellites: the United States placed the first such satellite in geostationary orbit at 35,880 kilometers above the equator. In this orbit, a satellite will maintain a fixed position related to a given point on the ground. The satellite receives signals from a ground station, boosts its power, and almost instantly retransmits the signal to another ground station. In this way, three properly positioned geostationary satellites can provide worldwide communication.

One year after the first American communications satellite was launched, the United States internationalized the technology, through the International Telecommunication Satellite Organization (Intelsat), "to achieve a global commercial telecommunications satellite system to provide, for the benefit of mankind, the most efficient and economical facilities possible, consistent with the best and most equitable use of the radio-frequency spectrum and orbital space." Founded by nineteen nations, Intelsat eventually gained well over a hundred members, and even included the Soviet Union in 1991—this after the latter attempted to create its own system (INTERSPUTNIK). When the USSR gave way to the Commonwealth of Independent States, the successor nations of Russia, Belarus, and Ukraine joined Intelsat. At the beginning, under U.S. sponsorship, Intelsat was very much an American organization, with the United States controlling 61 percent of the voting authority and all the technology. U.S. laws forbade sale of launch technology to Europe, and NASA was forbidden to provide launch service for satellites competing with Intelsat. Understandably, these restrictions were later relaxed. By 1993, Intelsat had fifteen satellites in orbit and carried roughly two-thirds of the world's overseas telecommunications traffic including telephone, telegraph, telex, television, data, and facsimile services. It carried out the principle in the Outer Space Treaty: "The . . . use of outer space . . . shall be carried out for the benefit and in the interests of all countries, irrespective of their degree of economic or scientific development, and shall be the province of all mankind."

The new network inspired the International Maritime Satellite Organization (Inmarsat), founded in 1976 for maritime communications, with sixtysome members by the early 1990s. The Soviet Union was a founding member of Inmarsat and, after creation of the CIS, its membership included Russia, Belarus, and Ukraine. Inmarsat "seeks to make provision for the space segment necessary for improving maritime communications and, as practicable, aeronautical communications, thereby assisting in improving communications for distress and safety of life, communications for air traffic services, the efficiency and management of ships and aircraft, maritime and aeronautical public correspondence services and radio-determination capabilities." As the above list of purposes shows, Inmarsat from the outset had the possibility of extending its purview well beyond maritime communications. In fact, 1985 amendments to its convention mandated it to provide global aeronautical communication services. In 1990, Inmarsat introduced commercial aeronautical services for airlines and corporate aircraft that currently include telephone, facsimile, telex, mail, data, position reporting, and fleet management, as well as distress and safety communications. It is also developing land mobile communications.

In 1971, the Soviets established the first domestic communications satellite system, enormously helpful because of their country's great territorial expanse, deploying four satellites in elliptical orbits—better than geostationary because of long linger times over the Northern Hemisphere. The CIS nations' continuation and expansion of the several domestic systems now operating over the former Soviet Union will obviously help make those countries more attractive to foreign investment.

After the Soviet Union established its first domestic satellite system, many other countries followed. Naturally, the next was Canada, again with a great physical expanse, followed by the United States, Japan, India, European countries collectively (Eutelsat) and individually, Indonesia, China, Mexico, Brazil, Australia, and members of the Arab League (Arabsat). There is also Eumetsat, the European Organization for Meteorological Satellites, established to ensure operation of Meteosat, the first European meteorological satellite, launched by ESA in 1977. Eumetsat usually has two Meteosats in orbit.

In recent years, the United States has developed navigation systems that are marvels of space science and make navigation an entirely new proposition compared to the complicated, handheld instruments of the distant past. One system, operated by the U.S. Navy, is the Navy Navigation Satellite System, which provides two-dimensional data (latitude and longitude). Established for military users, the system is open to civilian shipowners. In fact, 90 percent of users are civilian. The other system was used extensively by U.S. and allied forces in the Persian Gulf War of 1991 and is called the NAVSTAR Global Positioning System, or NAVSTAR GPS. It has a total of twenty-four satellites, of which four must be in view at one time, in order to provide three-dimensional data (latitude, longitude, and altitude), twenty-four hours a day, anywhere in the world, in all weather conditions. The system operates at two levels of accuracy: a "coarse" level telling users their approximate positions and a "precise" level within twenty centimeters.

Satellite observance of weather conditions around the globe has also become commonplace. In the past, weather predictors were of course little more than soothsayers, who if they did not engage in divination took measures no more scientific than to calculate prevailing winds. Now all that has changed. The first global cloud-cover picture was taken in 1960; three years later came automatic transmission of pictures, allowing real-time readout of local cloud pictures using an inexpensive ground station. The first spacecraft for the U.S. Weather Service went up in 1965, a spin-stabilized configuration with two television cameras, placed in sun-synchronous orbit. By 1970 it was possible to obtain day and night cloud-cover observations. Between 1975 and 1991 the United States launched twenty-nine weather observation satellites, not counting such satellites launched for the DOD.

From these U.S. beginnings, other nations developed their systems; the United States, ESA, and Japan now operate a worldwide network of high-altitude weather satellites that provides weather information for the rest of the world. Satellite-supplying participants help each other, as happened in a notable instance in February 1993. Ordinarily, the United States maintains two such weather satellites. One failed in 1989. The survivor, launched in 1987, had an intended five-year life. Its presumed replacement, the first of a new generation, was not ready because of schedule delays and budget overruns. So the United States "borrowed" a spare ESA satellite, *Meteosat-3,* which moved from a position over the Atlantic Coast of South America to the Pacific side. It was necessary, however, to delay the move until the United States built an $11 million communications relay station in Wallops, Virginia, for *Meteosat-3* was controlled by a station in Darmstadt, Germany.[41]

A graphic illustration of what weather observation satellites could do occurred in 1985 when a devastating cyclone approached Bangladesh. Fortunately, the Agency for International Development had financed and built a satellite weather-alert system, developed by NASA and operated by the National Oceanographic and Atmospheric Administration, the latter an agency of the U.S. Department of Commerce. In the past twenty years, Bangladesh, with 100 million people, had suffered an estimated 390,000 deaths from twenty-eight cyclones and other storms. In the cyclone of 1985, the death toll was 10,000, substantially less than would have died without the satellite system.[42]

The Land Remote-Sensing Commercialization Act of 1984 (Landsat Act) promotes commercial distribution of data from Landsat remote-sensing satellites, of major benefit in managing the Earth's natural resources. The secretary of commerce is responsible for this system and contracts with a private group, Earth Observation Satellite Company, for marketing the received data.

With the various satellite systems in orbit, finished or about to be, the world has become a far different place. The first communications satellite for Intelsat, *Early Bird,* had a capacity of 240 telephone circuits or one television channel. Each of the latest Intelsat-6 satellites can carry 24,000 telephone circuits as well as three television channels. By the end of the 1980s, the Intelsat system was being used by more than 150 nations, counting nonmembers, with 640 Earth stations connected by 2,259 transmission paths.[43] Communications have become "a whole new ballgame" as a result. Weather observation, navigation, and Earth resources monitoring have as well.

Meanwhile, as America was leading the world in development of satellite systems, a competitive launching arrangement was developing in Europe. The success, especially of Intelsat, irritated the French government of General de Gaulle, so they took the lead in challenging U.S. supremacy. The nations of Western Europe belonging to predecessor organizations of ESA agreed to a European launcher, *Ariane,* eventually handled under contract by a French company, Arianespace, with the purpose of putting satellites into geostationary orbit. *Ariane* made its last test flight in 1981 and was ready for business. With launches from French Guiana, capable of lifting two satellites into orbit, it soon proved its worth. When President Reagan removed commercial payloads from the shuttle, *Ariane* began to vie for American payloads. In addition, it launched scientific loads, such as a spacecraft named *Giotto,* which sailed off to Halley's Comet in

1985. It was the only one of five Halley's Comet probes to enter the nucleus of the comet, and it returned excellent data. The *Challenger* disaster in 1986, which destroyed an American shuttle and its entire crew, canceled the U.S. plan to observe the comet.[44]

Japan too sought to develop launch vehicles and satellites beginning in the 1970s, initially by persuading the United States to license the technology. Licenses carried limits less advanced than "state of the art." Japan also could not launch non-Japanese payloads without permission. Unlike the French, the Japanese found this arrangement satisfactory. Japan launched communications and weather satellites with American-supplied boosters, although the country has produced small boosters for science satellites. An issue arose between the two countries over unfair trading practices, namely, that Japan closed its communications satellite market to U.S. competition. In 1990, Japan agreed to allow foreign suppliers to bid for operational satellites, and a contract went to Space Systems/Loral (formerly Ford Aerospace).

India and Israel have the capacity to enter the commercial satellite market but thus far have made few space launches.

China entered the competition with its first commercial launch in 1987 (China had launched its first military satellite in 1970), taking a materials-processing experiment into space for a French company. The Chinese arranged a similar flight in 1988 for Intospace, a German consortium. A third was a communications satellite for Asiasat, Inc., a company based in Hong Kong. Following the Tiananmen Square uprising, a question arose over whether to allow Chinese vehicles to launch U.S. satellites. In 1991, the Chinese launched only a single communications satellite, designated for geostationary orbit, but a failure of the *Long March-3*'s third-stage rocket left it in the wrong orbit. Meanwhile, other contretemps were making China's role in commercial launches difficult. Because the difference between civil and military rockets is minuscule, the United States was sensitive to any export of Chinese rockets. In 1991, President Bush announced that it would be inappropriate to approve further export licenses for satellite components launched by China. Technically speaking, the exports covered by the Missile Technology Control Regime (MTCR) did not include satellite components, only rockets—but the Chinese were proposing to launch a Swedish satellite with U.S. components, and this fact gave the president leverage. When the Chinese agreed to adhere to the MTCR, the president lifted the sanctions. Another difficulty with China concerned an agreement signed in 1989 that China would charge prices "on a par" with Western launch-service providers. The Chinese contracted with the

Arab Satellite Consortium for a $25-million launch, but France objected because of the low price. ESA's Arianespace later received the contract. Then the perennial human rights difficulties with China again intervened against Chinese commercial launches. China had contracts to launch three foreign satellites in 1992: two for Australia and one for Sweden. But in the spring of 1991, Indonesia chose an American company, McDonnell Douglas, to launch its *Palapa* satellite instead of selecting the China Great Wall Industry Corporation, probably because of uncertainty surrounding U.S. export of components for satellites launched by China.

After the Soviet Union broke up, the CIS nations gave evidence of entering the space business, perhaps in an indirect way. An ingenious proposal came from an Australian company, Cape York Space Agency, to avoid technology transfer problems by buying Zenit rockets and sending them up from Cape York, Australia. This would circumvent the U.S. ban on export to the CIS countries of satellites containing U.S. components. The agency would own the rockets, and an Australian or American company (perhaps U.S. Space Boosters, Inc., a division of United Technologies) could put the satellites on the rockets, so no Russians needed to go near the launch site. U.S. companies in the commercial launch business—McDonnell Douglas, General Dynamics, and Martin Marietta—objected to U.S. Space Boosters' application for a technical assistance agreement from the State Department, claiming it would permit CIS (and Chinese) launch-service providers to lower prices and undercut the launch market. The Cape York Space Agency went bankrupt, but another Australian company, Space Transportation Systems, attempted to proceed with the idea.

Another scheme to bring in the Russians surfaced in 1991. The proposal was to sell military satellites, weighing five hundred pounds, launched six at a time on a single Soviet Tsyklon booster. The hope was first to have a set of commercial versions in orbit, then a twenty-four spacecraft constellation. An American support group appeared, the Consortium of Small Satellite Constructors and Service Providers, based in Warwick, New York. By having the Soviets build and launch the system, the group would avoid U.S. licensing requirements. It all sounded like a suitably capitalist enterprise. The satellites were designed by NPO Precision Instruments of Moscow and built by NPO Applied Mechanics of Krasnoyarsk.[45]

Conclusion

And so, in less than half a century, the space age has passed from its beginnings into something resembling—one uses the word gingerly—maturity. In no single aspect of human activity since October 1957 has there been so many changes, so many accomplishments, as in the science and technology of space.

Perhaps the most important change was the marked shift from U.S.-Soviet competition to widespread international cooperation. To be sure, in NASA's statutory statement of 1958 a mandate appeared for international cooperation: "The Congress hereby declares that it is the policy of the United States that activities in space should be devoted to peaceful purposes for the benefit of all mankind." And President Kennedy asked the nations to do the big things together. Competition, however, occurred if only because of the military implications of space science and because of American public opinion—the latter being certain that the Russians were ahead. On their part, Soviet leaders did their best to instill among Western audiences a belief that the United States and the Soviet Union were racing for supremacy in space and that the Soviets were ahead. Throughout the late 1950s, this was the nature of the space age; such also was its nature into the 1960s when the Soviets tried to beat the Americans to the Moon and failed only because they could not produce boosters capable of lifting their space capsules.

In the mid-1960s, competition began to shift toward cooperation, for Soviet leaders realized that their space scientists could not make the boosters they needed; at about the same time, U.S. leadership with Intelsat showed the way toward an international cooperation that would transform communications on land and sea, bringing the world together in ways that all the heralded revolutions in communications of the nineteenth century—telegraphs, telephones, steam navigation—failed to accomplish. It is a curious but now obvious truth that at the height of the cold war, during the Kennedy-Khrushchev confrontation over Cuba and the Soviet race to obtain equality in intercontinental ballistic missiles—a race that achieved what Henry A. Kissinger in 1969 described as "sufficiency"—two turns of events in space science pointed to cooperation. Thereafter, with increasing internationalization of the science and technology and now with the end of the cold war and collapse of the Soviet Union, international cooperation has become ever more evident.

It has been a long passage from the 1950s to the 1990s, pursued on the U.S. side largely by the nation's presidents, without much understanding by the American people, pursued at the beginning and perhaps through most of its passage for the wrong or at least questionable reasons. But in its results, the cornucopia of achievements, it has been vastly worthwhile.

Notes

1. According to Yuli B. Khariton, a leading participant in the Soviet nuclear program from its inception, the Soviets were well on their way toward making their own A-bomb when they received the full plans of the Nagasaki bomb. Stalin advised using the American plans, remarking, "We have to move broadly, with Russian sweep." In 1951, the Soviets detonated their own version. See *New York Times,* Jan. 14, 1993. Khariton, age 88 in 1993, also related that Andrei D. Sakharov produced an H-bomb in 1953. All the evidence on the latter point, however, is to the contrary. The Soviet test device in 1953 was not of an H-bomb but an enhanced A-bomb, wrapped in layers of the compound lithium deuterride, equal at most to a megaton of TNT. See William J. Broad, "Soviets Shown to Have Lagged on H-Bomb in 50's," *New York Times,* Oct. 7, 1990. In offering this assertion about Sakharov, Khariton raised an interesting possibility that, if true, would underline Soviet nuclear weakness. Khariton said that, just before the Soviet test in 1953, Stalin asked the physicist who directed the Soviet nuclear program, Ioor V. Kurchatov, whether there was enough plutonium to build both a test devi.ce and a similar bomb to keep "in reserve." Kurchatov replied that there was only enough for a test, implying that in 1949–53 the Soviet nuclear arsenal was very small.

2. Herbert York, *Race to Oblivion: A Participant's View of the Arms Race* (New York: Putnam, 1970).

3. The leading general account is Walter A. McDougall, . . . *The Heavens and the Earth: A Political History of the Space Age* (New York: Basic Books, 1985).

4. R. Cargill Hall, "The Origins of U.S. Space Policy: Eisenhower, Open Skies, and Freedom of Space" (unpublished manuscript), pp. 24–25, NASA Historical Reference Collection, NASA History Office, Washington, D.C.

5. Ibid., p. 9.

6. "The IGY scientific satellite program was clearly identified as a stalking horse to establish the precedent of overflight in space for the eventual operation of its military alternate" (ibid., p. 23). See also McDougall, . . . *The Heavens and the Earth,* pp. 110, 194.

7. McDougall, . . . *The Heavens and the Earth,* p. 258.

8. According to Johnson, "That sky had always been so friendly, and had brought us beautiful stars and moonlight and comfort and pleasure; all at once it seemed to have some question marks all over it because of this new development. I guess for the first time I started to realize that this country of mine might perhaps not be ahead in everything." Quoted in John M. Logsdon, *The Decision to Go to the Moon: Project Apollo and the National Interest* (Cambridge, Mass.: MIT Press, 1970), p. 21. For other Johnsonian statements on the space program as well as his power plays, see Robert A. Divine, "Lyndon B. Johnson and the Politics of Space," in *The Johnson Years,* vol. 2, *Vietnam, the Environment, and Science,* ed. Robert A. Divine (Lawrence: University Press of Kansas, 1987), pp. 218–53. Perhaps his most telling remark concerned the Moon race: "I do not believe that this

generation of Americans is willing to resign itself to going to bed each night by the light of a Communist moon" (p. 234). Johnson capitalized on Reedy's advice and, among other efforts, worked through Senator Richard B. Russell Jr. of Georgia to get one of his presidential rivals in the Democratic party, Senator Stuart Symington of Missouri, out of the *Sputnik* investigation. The Democratic lawyer, James H. Rowe Jr., also encouraged Johnson to urge the space race (p. 220). Johnson chaired the military preparedness subcommittee hearings. He was not always present, and perhaps some of his overstatements may be credited to that fact, but he clearly knew more than Kennedy about the real situation.

9. Logsdon, *Decision to Go to the Moon*, recounts the decision to "go." Charles Murray and Catherine Bly Cox, *Apollo: The Race to the Moon* (New York: Simon and Schuster, 1989), provide an excellent overview of a complex subject. NASA official histories of the subject include: Charles D. Benson and William Barnaby Faherty, *Moonport: A History of Apollo Launch Facilities and Operations* (Washington, D.C.: NASA [SP-4204], 1978); Courtney G. Brooks, James M. Grimwood, and Lloyd S. Swenson Jr., *Chariots for Apollo: A History of Manned Lunar Spacecraft* (Washington, D.C.: NASA [SP-4205], 1979); Roger E. Bilstein, *Stages to Saturn: A Technological History of the Apollo/Saturn Launch Vehicles* (Washington, D.C.: NASA [SP-4206], 1980); Arnold S. Levine, *Managing NASA in the Apollo Era* (Washington, D.C.: NASA [SP-4102], 1982); W. David Compton, *Where No Man Has Gone Before: A History of Apollo Lunar Exploration Missions* (Washington, D.C.: NASA [SP-4214], 1989); Sylvia D. Fries, *NASA Engineers and the Age of Apollo* (Washington, D.C.: NASA [SP-4104], 1992).

10. "We were eating their lunch, eating their lunch." Robert F. Freitag, interview by Sylvia D. Kraemer, May 16, 1985, p. 31, NASA Historical Reference Collection.

11. Cost figures are in a letter from James Webb to John L. Sloop, June 14, 1972, Si Info 9, Logsdon case file, NASA Historical Reference Collection.

12. Arnold W. Frutkin, interview by Eugene M. Emme and Alex Roland, Apr. 4, 1974, pp. 28–29, NASA Historical Reference Collection. In an earlier interview, Frutkin noted, "There was just no practical basis for giving any thought whatever to international cooperation in that program. I would argue that strenuously with anyone who thought that—I mean, given the premises of the Apollo program, it was absurd at that time to think of international cooperation, and so a very large segment of the program really went ahead without international involvement. . . . It is also a very important thing to know that throughout the sixties, throughout the entire term of the sixties, no foreign government, no foreign official, ever hinted of a desire to enter the manned spaceflight program anyway, so there was no ready basis to build on." Frutkin, interview by Logsdon, July 30, 1970, pp. 17–18, NASA Historical Reference Collection. See also Frutkin, *International Cooperation in Space* (Englewood Cliffs, N.J.: Prentice-Hall, 1965).

13. For the above quotations, see Ken Hechler, *Toward the Endless Frontier: History of the [House] Committee on Science and Technology, 1959–79* (Washington, D.C.: Government Printing Office, 1980), pp. 394–95.

14. See Edward Clinton Ezell and Linda Neuman Ezell, *The Partnership: A History of the Apollo-Soyuz Test Project* (Washington, D.C.: NASA [SP-4209], 1978).

15. For an excellent survey of the literature, see Roger D. Launius and Aaron K. Gillette, comps., *Toward a History of the Space Shuttle: An Annotated Bibliography* (Washington, D.C.: NASA, 1992). This is the first volume in Studies in Aerospace History, sponsored by the NASA History Office.

16. Howard E. McCurdy, *The Space Station Decision: Incremental Politics and Technological Choice* (Baltimore: Johns Hopkins University Press, 1990), p. 72. McCurdy's book is an ideal combination of technical writing and literary ability. See also McCurdy, *Inside NASA: High Technology and Organizational Change in the American Space Program* (Baltimore: Johns Hopkins University Press, 1993).

17. Douglas R. Lord, *Spacelab: An International Success Story* (Washington, D.C.: NASA, 1987), p. 222.

18. McCurdy, *Space Station Decision*, p. 200.

19. Logsdon, "U.S.-European Cooperation in Space Science: A 25-Year Perspective," *Science*, Jan. 6, 1984, p. 14. See also the amusing explanation in Adam Gruen, "Partners," pp. 92–93: "In effect, this left European nations to foot the bill for completing, storing, or junking their half of the program." Gruen's chapter is part of a manuscript on the space station that takes up where McCurdy's book stops. Copy in NASA Historical Reference Collection.

20. Lord, *Spacelab*, p. 274.

21. McCurdy, *Space Station Decision*, p. 99.

22. *Space News*, June 17, 1991; McCurdy, *Space Station Decision*, p. 197. For this and subsequent references to aerospace periodicals, also most newspapers, I am indebted to Lee D. Saegesser, archivist of the NASA History Office, who has systematically searched the periodicals and newspapers and clipped and filed anything bearing on NASA.

23. McCurdy, *Space Station Decision*, p. 202.

24. Logsdon, *Together in Orbit: The Origins of International Participation in Space Station Freedom* (Washington, D.C.: George Washington University Press, 1991), pp. 37, 129–30.

25. See McCurdy, *Space Station Decision*, pp. 168–69, 171, 174, 232 (Beggs quotations are on p. 168). McCurdy concluded (pp. 234–35) that to make the station politically acceptable, and the shuttle before it, "to preserve flexibility and save money, politicians managed to avoid a commitment to a long-range space policy for twenty years." The cost, he wrote, to people who had to carry projects forward, was very high. Use of "incremental strategies" was especially evident in the space station decision.

26. *Christian Science Monitor*, Oct. 21, 1992.

27. McCurdy, *Space Station Decision*, p. 171.

28. Logsdon, *Together in Orbit*, pp. 48–53. "The issue of international participation was not separately raised with President Reagan; his approval came in the form of overall approval of the speech text" (p. 53). A year later, Freitag of-

fered his own explanation of why the president supported the shuttle: "I think . . . he was a Boy Scout at heart. I think he had always been interested in space and this was nothing new. And that the whole shuttle enterprise was thrilling to him. . . . People like Beggs were deeply influential with him. People like the Congress committee chairmen. The other thing about it, it was the only good answer. What other answer was there? . . . Assuming you wanted a big project, assuming you wanted something that had all the virtues of maintaining NASA as a viable organization, yet serving a tremendous amount of apparent usefulness, there was no alternative. You can't build another shuttle this time; a lot of nickel-and-dime projects are interesting but . . . a mission to the moon is not going to grab it. There are no science projects of that magnitude. A telescope is big, but not that big. Mars is a long ways away. SDI might have been interesting from the standpoint of the defense side, but not on the civil side. So I think there is a certain lack of alternatives to it, too." Freitag, interview by Kraemer, July 12, 1985, pp. 19–20, NASA Historical Reference Collection.

29. Logsdon, *Together in Orbit,* p. 62.

30. McCurdy, *Space Station Decision,* p. 199, points out a domestic awkwardness of the proposed Canadian space-station arm. It had to compete with a proposal for a radar satellite providing all-weather monitoring of forest fires and sealanes in remote regions, together with a communications satellite system to link mobile radio users in outlying areas with the rest of Canada. There was only enough money for one project.

31. Logsdon, "International Cooperation in the Space Station Program: Assessing the Experience to Date," *Space Policy* 7 (Feb. 1991): 38, 41–44. For its special subject, see Nandasiri Jasentuliyana, ed., *Space Law: Development and Scope* (Westport, Conn.: International Institute of Space Law, 1992).

32. *Space News,* Mar. 18, 1991.

33. *Washington Post,* Dec. 6, 1991; *Aviation Week and Space Technology,* Feb. 25, Mar. 11, 1991, and Dec. 7, 1992.

34. *New York Times,* Mar. 12, 1993.

35. Editorial, *New York Times,* May 15, 1992. It proved exceedingly difficult to attach the capture bar to the satellite, an apparently simple task for which the astronauts had trained for two years in an extremely expensive pool designed for such a purpose and recently finished at the Kennedy Space Center. The task was to get the satellite into the bay of the shuttle, attach a rocket, and send the satellite into its correct orbit. The jockeying of the satellite also raised questions about assembling a full-scale space station, with thousands of struts and components, sent up piece by piece in shuttles.

36. Shuttle flight costs are a matter of debate. NASA has estimated a flight at $44 million, the "marginal cost" of adding a flight to the ongoing shuttle program— for extra fuel and salaries. It calculated the "average cost" of a flight at $414 million. According to an outside team of academicians, if the expense of repair and launching is added, the cost per flight is $1.1 billion. Including costs of develop-

ment and improvement jumped the figure to $1.7 billion. See *New York Times,* Mar. 22, 1993.

37. McCurdy, *Space Station Decision,* pp. 73, 88; *Space News,* Mar. 30, Sept. 28, Nov. 16, 1992; *New York Times,* Aug. 23, 1992; *Aerospace America,* Oct. 1992, p. 31. Thirty years ago, Kenneth S. Pedersen has suggested, it would have seemed impossible to think of going to the Moon or Mars aboard a "Titangia." Now (albeit with some new name) it just might happen; see "Thoughts on International Space Cooperation and Interests in the Post–Cold War World," *Space Policy* 8 (Aug. 1992): 217.

38. *Space News,* Sept. 28, 1992. In 1971, the Soviets launched *Salyut,* the first in a series of small orbiting laboratories, and took a conservative approach over the next fifteen years. Not until 1986 did they launch a module with a multiple docking port named *Mir.* They brought the crew back to Earth before linking the first laboratory module. At no time did they attempt a permanently manned station. The Americans, however, "did not want to spend fifteen years plodding along with a small can in the sky." See McCurdy, *Space Station Decision,* pp. 108–9.

39. *Christian Science Monitor,* Oct. 21, 1992; *Aerospace Daily,* Nov. 25, 1992; *Aviation Week and Space Technology,* Dec. 7, 1992; *Space News Roundup* (Johnson Space Center), Dec. 18, 1992, p. 4; Liz Tucci, "Agreement Blesses Soyuz Use as NASA Rescue Vehicle," *Space News,* Dec. 14, 1992, p. 4. The lifeboat issue has a side to it not often discussed. A former leading official in the construction of space station *Freedom* told me that the reason he and his colleagues made no provision for a lifeboat was that in the event of a disaster to the station no lifeboat would be needed—crew members would already have perished. Presumably for political reasons, this point is impossible to make publicly.

40. The following account is indebted to Marcia S. Smith, *Space Activities of the United States, Soviet Union, and Other Launching Countries/Organizations: 1957–1991* (Washington, D.C.: Congressional Research Service of the Library of Congress, 1992). Smith is the library's specialist in science and technology policy in the division of science policy research. See also *26 Years of NASA International Programs* (Washington, D.C.: NASA, 1984). As of January 1, 1984, NASA had engaged in 38 cooperative spacecraft projects, 1,774 cooperative sounding-rocket projects, 163 cooperative ground-based projects for remote sensing, and 95 reimbursable launchings of non-U.S. spacecraft and was operating 48 overseas tracking stations in twenty countries. Resident research associateships had reached 1,412, and NASA had received 85,177 foreign visitors. More recent figures are not available.

41. *New York Times,* Feb. 25, 1993.

42. *Washington Times,* May 29, 1985.

43. Jonathan F. Galloway, "Space Law in the United States," in *Space Law,* ed. Jasentuliyana, p. 78.

44. The arrangement for *Ariane* was closely linked with the agreement to go ahead with the ESA contribution to the American shuttle program; three days

before signing an MOU with NASA on *Spacelab,* the European countries decided to sign up for *Ariane.* Somewhat covered in a "package deal," the Europeans agreed to develop the maritime communications network and place all their enterprises under ESA. See Logsdon, *Together in Orbit,* p. 11.

45. In addition to the U.S. marketing consortium, another group examining the spacecraft was SatelLife, a supplier of health-related information to developing nations (*Aviation Week and Space Technology,* Mar. 25, 1991).

7

National Leadership and Presidential Power

John M. Logsdon

Many benefits have been claimed for the space program—scientific discovery, economic growth, technological advancement, job maintenance, and educational excellence. None of these benefits alone justify the direct presidential involvement often sought for this endeavor. The other essays in this book describe in detail the importance of presidential leadership in space policy formulation and execution; my task here is to explore how presidents themselves have defined the concept of leadership. In essence, a primary determinant of a president's overall attitude toward space has been the judgment of the program's usefulness as an instrument for projecting an image of U.S. leadership to a global community and as a means of attaining a range of domestic goals.

Leadership

The quest for leadership has been a central feature of U.S. space policy from the very beginning. The bill that the Eisenhower administration sent to Congress in April 1958, which became the Space Act of 1958, set as a policy objective "the preservation of the role of the United States as a leader in aeronautical and space science and technology."[1] Dwight D. Eisenhower's objective is notable in its use of the article "a" as opposed to "the." In other words, the president's bill did not ask Congress for a formal statement of policy that called for the United States to be *the* leader in space science and technology but rather to be *a* leader.

Even so, over the past four decades, the word "leadership" has been pervasive in most discussions and reports dealing with space, to the point that it almost loses any specific meaning and becomes only expected rhetoric. In fact, there are many possible meanings of leadership. Some have to do with relative standing in a competition, identifying who is "ahead" by some measure. Other definitions describe the absolute character or quality of a country's efforts; in this case, there can be many leaders. But beneath the rhetoric is a basic understanding—that being perceived by others, and by itself, as being at the forefront of space capabilities and achievements is in the American national interest.

The United States has actively sought space leadership when the sitting president has made the judgment that such leadership was an important element of U.S. power. In this context, U.S. power is defined as the ability to influence events and choices around the world so that they are congruent with American interests. It is a little bit unfashionable to talk about power and its use, and members of the space community do not usually link space activities to the creation and use of national power. The desire for power in international relations can be seen, however, as the fundamental underpinning of the presidential quest for space leadership. A brief and selective comparative overview of how various presidents have assessed the links among the concept of leadership, the U.S. civilian space program, national power, and international prestige is instructive in this regard.

Historical Overview

Dwight D. Eisenhower was never convinced that being the leader in space was important to U.S. international interests. He certainly heard arguments to the contrary, but he used the power of his presidential office to reign in advocates of an expensive and ambitious space program. As early as May 1955, he approved a space policy statement, National Security Council (NSC) 5520, that warned:

> Considerable prestige and psychological benefits will accrue to the nation which first is successful in launching a satellite. The inference of such a demonstration of advanced technology and its unmistakable relationship to intercontinental ballistic missile technology might have important repercussions on the political determination of free world countries to resist Communist threats, especially if the USSR were to be first to establish a satellite.[2]

Eisenhower's resistance to the idea that the launch of a small satellite had any real significance to the relative military standing of the United States and the Soviet Union has been well documented by several historians, and the essay in this volume by David Callahan and Fred I. Greenstein makes the point well. The phrase "calm conservatism"[3] has been used to describe the president's stance as others pushed for early U.S. entry into a prestige-oriented space race with the Soviet Union.[4] Even so, Eisenhower approved two post-*Sputnik* space policy statements that made the link between space leadership and U.S. global interests. For example, the first post-*Sputnik* statement of U.S. space policy, NSC 5814/1 of August 1958, noted that "to be strong and bold in space technology will enhance the prestige of the United States among the peoples of the world and create added confidence in U.S. scientific, technological, industrial and military strength." This statement also called for developing the space capabilities needed "to establish the United States as a recognized leader in space."[5]

Nonetheless, Eisenhower was unwilling to invest U.S. resources in space primarily on leadership grounds; rather, there had to be other tangible benefits for a project to be approved. Eisenhower acknowledged the space race but was unwilling to pay the price of a clearly leading position in that competition. His attitude is perhaps best captured in an excerpt from his January 1960 space policy statement:

> To minimize the psychological advantages which the USSR has acquired as a result of space accomplishments, *select from among those current or projected U.S. space activities of intrinsic military, scientific or technological value*, one or more projects which offer promise of obtaining a demonstrable effective advantage over the Soviets and, so far as is consistent with solid achievements in the over-all space program, stress these projects in present and future programming [emphasis added].[6]

In stark contrast, John F. Kennedy was clearly convinced of the link between space leadership and core U.S. interests, and that conviction led to his decision to use his public office to mobilize the national will and resources that produced Project Apollo. The decision to go to the Moon has been described by Walter A. McDougall as an "overdetermined event";[7] certainly, the record on why Kennedy approved Apollo is very clear, and the quest for leadership was at the center of that decision. On April 20, 1961, eight days after the Soviet launch of the first human, Yuri Gagarin, into orbit, Kennedy asked Vice President Lyndon B. Johnson to conduct a survey of where the United States stood in space. In particular, Kennedy asked, "Do we have a chance of beating the Soviets by putting a laboratory in space, or by a trip around the moon, or by a rocket to go to the moon and back with a man? Is there any other space program that promises dramatic results in which we could win?"[8]

Johnson's reply was equally clear. The vice president on May 8 transmitted to the president a report, signed by NASA Administrator James Webb and Secretary of Defense Robert McNamara, that represented a radical change in policy from that of the Eisenhower administration. Rather

208 John M. Logsdon

than setting the terms for competition with the Soviet Union so that projects must have other elements of "intrinsic merit," the report recognized that "major successes, such as orbiting a man as the Soviets have just done, lend national prestige even though the scientific, commercial or military value of the undertaking may by ordinary standards be marginal or economically unjustified. . . . *This nation needs to make a positive decision to pursue space projects aimed at enhancing national prestige."*[9]

What is less clear is the consistency of Kennedy's view of the linkage of space achievement, U.S. leadership, and international realities. Project Apollo, for Kennedy, may have been more a response to the specific political situation in the first months of his presidency than a symbol of his long-term commitment to U.S. space leadership. In both 1962 and 1963, Kennedy called for major, highly classified reviews of the space program to determine, among other issues, whether to go ahead with Apollo as planned.

Then, on September 20, 1963, speaking to the General Assembly of the United Nations, Kennedy suggested "a joint expedition to the moon. . . . Why, therefore, should man's first flight to the moon be a matter of national competition?"[10] This was not a casual remark or speechwriter's rhetoric; rather, it indicated that serious consideration was being given by Kennedy and his advisors to recasting Apollo as a cooperative project. Just ten days before he was assassinated, Kennedy issued a National Security Action Memorandum directing Webb to take the "initiative and central responsibility within the Government for the development of a program of substantive cooperation with the Soviet Union in the field of outer space." That program was to include "cooperation in lunar landing programs."[11]

If one were to give full credence to Kennedy's public rhetoric, he above all U.S. presidents was committed to placing the force of his office behind the goal of U.S. space leadership. This led to the so-called "Apollo syndrome" of NASA, a belief that a president by his mere announcement of bold space programs can ensure their support and completion. In private—and this was something most advocates of a strong space program did not know and have begun to perceive only within the last decade—Kennedy's approach seems to have been somewhat different, as he wrestled with the reasonableness of what his public rhetoric had set in motion. In remarks the day before he was assassinated, Kennedy said, "The space program stands on its own as a contribution to national strength. . . . I think the United States should be a leader . . . should be second to none."[12] Whether or not Kennedy fully subscribed to this view, history is likely to record him as the most pro-space president of the twentieth century. Certainly, supporters of space exploration have already assigned him that distinction.

As senator and vice president, Johnson seems to have been a staunch supporter of the space program and a firm believer in the importance of space leadership to U.S. national interests. He played a key role in developing the recommendations that led to Apollo and in assembling the political coalition in support of the undertaking. When Kennedy asked him in 1963 to identify the benefits that could flow from the space program, Johnson, in a May 13 response, noted that, in addition to whatever tangible benefits might result, "a much more fundamental issue is at stake—whether a dimension that can well dominate history for the next few centuries will be devoted to the social system of freedom or controlled by the social system of communism. . . . Our space program has an overriding urgency that cannot be calculated solely in terms of industrial, scientific, or military development. The future of society is at stake."[13]

Just over six months later, Johnson was president. Robert Dallek's chapter in this volume quotes him as recalling, "I think I spent more time in the space field . . . up to '63 than I did after I became President."[14] While Johnson's views regarding the importance of space were unlikely to have changed drastically, other concerns—financing the Great Society, urban unrest, and the draining conflict in Southeast Asia—dominated his presidency.

Johnson remained committed to finishing Apollo but, much to the frustration of NASA Administrator Webb, he was unwilling to approve new projects that would build on the capabilities developed during the 1961–65 space mobilization. After peaking in 1965, NASA's budget for space began a rapid downslide that continued for the next decade. Johnson may have believed in the importance of space leadership, but he found himself unable to allocate to the space program the resources required to sustain that leadership once America reached the Moon. His support for space is unlikely to be recorded as one of the highlights of his years in the White House. At the same time, the combination of his support for Project Apollo and his unwillingness to commit to large follow-on projects reinforced the image of the president as omnipotent with respect to the broad outlines of U.S. space policy.

By the time Richard Nixon assumed the presidency in January 1969, the NASA budget had been reduced to approximately $4 billion from its $5.2 billion peak. Nixon's space transition team, headed by the Nobel Prize winner Charles Townes, told the new president that a budget at the $4 billion level was "adequate" and that at that budget level the United States could carry out "an adequately competitive space effort." The team believed it was unnecessary to compete with the Soviet Union "in detail" but that "the U.S. effort must be as strong overall as that of the Soviet Union."[15]

Nixon seems to have taken to heart this less-than-ringing call to continued space leadership. Like everyone else, he basked in the glory of the first lunar landing in the summer of 1969. But when confronted by the bullish recommendations of the Space Task Group in September 1969, Nixon remained silent for six months, then announced in March 1970 that "our approach to space must be bold—but it must also be balanced. . . . Space expenditures must take their place within a rigorous system of national priorities."[16]

Reducing the priority and the budget demands of the space program after the first lunar landing was a conscious decision on Nixon's part. He was advised that a majority of "the heart of your constituency" believed that less money should be spent on space.[17] After two years of reductions, however, one of the president's most trusted advisors told him that the cuts were going too far and that NASA was being proposed for further reductions "because it is cuttable, not because it is doing a bad job or an unnecessary one."[18] Caspar Weinberger, then deputy director of the Office of Management and Budget (OMB), couched his argument against additional cuts in the NASA budget in terms of the space program's links to the U.S. image in the world:

> It would be confirming, in some respects, a belief that I fear is gaining credence at home and abroad: That our best years are behind us, that we are turning inward, reducing our defense commitments, and voluntarily starting to give up our super-power status, and our desire to maintain our world superiority.
>
> America should be able to afford something besides increased welfare, programs to repair our cities, or Appalachian relief and the like.[19]

Nixon, with his intense sense of world affairs, was impressed by Weinberger's argument. At the top of the memo he penned, "I agree with Cap." NASA's budget slide began to bottom out, and within six months the president approved development of the space shuttle.

An important change in the way the United States would pursue space leadership occurred during the Nixon presidency. With the successful *Apollo 11* mission, the United States had clearly won the space race. Now the country would seek to expand its cooperative efforts with both its allies and the Soviet Union. George Low, deputy administrator of NASA, recorded Nixon's views after a January 1972 meeting with him: "The President said that he is most interested in making the space program a truly international program and that he had previously expressed that interest.

He wanted us to stress international cooperation and participation for *all* nations. . . . He is not only interested in flying foreign astronauts, but also in other types of meaningful participation, both in experiments and even in space hardware development."[20] By demonstrating the quality of its capabilities within a cooperative project, the United States could continue as the leader even as nations began to work together in space. This is a theme in U.S. space policy that continues even now, albeit with some rough spots along the way.

There is little specific to link President Gerald Ford to major space policy or program decisions. Indeed, in most areas the two and a half years of Ford's presidency were marked by continuity with the policy directions established during the Nixon administration. Space was no exception. Thus, no major decisions on space were addressed by the White House while Ford was president, and he took no major initiatives. The NASA budget did begin a gradual upward trend, and in his last budget Ford approved "new starts" for both the *Galileo* mission to Jupiter and the Hubble Space Telescope. This is the most identifiable mark on space policy made by the Ford administration, since approving two major space science programs at the same time had not happened before 1977 and certainly has not happened since.

The tone of Jimmy Carter's approach to space may have been set by the advice given in his transition team's report on NASA, which was prepared by Nick MacNeil. That report noted, with a high degree of skepticism, that a major argument in support of NASA was that "to keep our political and cultural values in high esteem, we must periodically give a display of technological virtuosity. . . . We transfer vigor and Number 1 status in a particular field, to the nation as a whole. Selling international prestige on this basis panders to people's insecurities." MacNeil suggested that the Carter administration "separate the sales pitches that involve international prestige, displays of power, [and] Buck Rogers entertainment. These play up to our insecurities and offer satisfactions and diversions that are artificial."[21]

While these views may not have had a direct impact on Carter's attitude toward space, he certainly did not embrace the concept that space leadership was important to core U.S. interests. The White House issued a statement of national space policy in October 1978 that took a quite measured view, noting that "space activities will be pursued because they can be uniquely or more efficiently accomplished in space. Our space policy will become more evolutionary rather than centering around a single, massive engineering feat. Pluralistic objectives and needs of our society will set the course for future space efforts."[22]

The same statement listed the fourth of nine objectives of the civilian space program as being to "assure U.S. scientific and technological leadership for the security and welfare of the nation."[23] Like the switch in emphasis from competition to cooperation under Nixon, this statement appears to mark a watershed in U.S. space policy. Rather than seeking leadership through achievement, now the United States would lead in technology and capability, without committing itself to their visible use.

An additional clue to Carter's attitude toward the space program can be found in his Rose Garden remarks on the occasion of the tenth anniversary of the first lunar landing. After accepting mementos from *Apollo 11* astronaut Neil Armstrong and NASA Administrator Robert Frosch, the president noted, "We landed on the Moon because our Nation set a firm goal, and we united behind that effort." He went on to suggest that other priorities were more important in his view than space achievement: "Today, we face an equally challenging goal in fighting for energy security. . . . We will win energy security for our Nation in the same way we won the race to the Moon."[24] These were hardly the words of a president committed to U.S. leadership in space.

If his rhetoric were taken at face value, President Ronald Reagan was a quite strong supporter of U.S. leadership in space. But, as noted in Lyn Ragsdale's chapter in this volume, space during the Reagan presidency was not a top priority issue—at least not before the *Challenger* accident. In reality, Reagan gave the civilian space program just enough priority and budgetary support to allow it to move forward at a modest pace, but he certainly did not claim it as a major element of his presidential strategy. Reagan approved a major new initiative, the space station, and he used it as a highly visible symbol of cooperation among the United States and its "friends and allies."[25] But in this arena the chinks in the armor of the "imperial presidency" began to be noticeable. The Reagan administration could secure only limited funds for the station from Congress.

The Reagan space transition team was composed primarily of veterans of the space program and led by the longtime NASA official George Low. Not surprisingly, particularly in contrast to the Carter transition document, the team's report was positive with respect to the space program's values. It noted that "national prestige is how others view us, the global perception of this country's intellectual, scientific, technological, and organizational capabilities. In recent history, the space program has been the unique positive factor in this regard."[26]

The first overall statement of space policy by the Reagan administration was issued on July 4, 1982, as the president witnessed a landing of the shuttle *Columbia* at Edwards Air Force Base. As Ragsdale notes, the event was staged as a patriotic festival. Second only to strengthening "the security of the United States" as an articulated policy goal was maintaining U.S. "space leadership."[27] Apparently, the goal of leadership was to have higher priority under Reagan than it had for his predecessor. But precisely what that meant in practice was not clear.

One clue to the meaning (or lack thereof) of the term "leadership" in the thinking of the Reagan administration is its "National Space Strategy," which the president approved on August 15, 1984. The strategy states:

—The STS [Space Transportation System, i.e., space shuttle] is a critical factor in maintaining U.S. space leadership.

—The development of a civil Space Station will further the goals of space leadership. . . .

—Major long-range goals for the civil space program are essential to meeting the national commitment to maintain United States leadership in space. . . .

—The U.S. civil space science program is an essential element of U.S. leadership in space.[28]

In other words, by 1984, essentially everything the United States was doing in space was linked, at least rhetorically, to the quest for leadership. By 1988, the final Reagan-era statement of space policy said that "a fundamental objective guiding United States space activities has been, and continues to be, space leadership."[29] It is not clear that "leadership" retained much specific content at this point.

Possibly a more precise clue to what issues related to space were most important to Reagan can be gleaned from the highly polished NASA sales pitch for the space station that was presented to the president on December 1, 1983. Given that it certainly was in NASA's interest to stress those themes most likely to elicit a positive response from the president, the agency chose to highlight, among other things, that a U.S. space station would "implement the overriding theme of your space policy: U.S. leadership in space."[30]

It is fair to say that the desire for leadership was an important factor in shaping Reagan's attitude toward space, but overall he did not place top priority on a fast-paced U.S. space effort. Efforts to resurrect the notion of a "space station race" with the Soviet Union were not notably successful. By the end of his presidency, Reagan was talking about space cooperation with our former adversary as well as with traditional U.S. friends and allies.

Preeminence

Preeminence differs from leadership as a goal of U.S. space policy. First, no president has uttered publicly the word preeminence when talking about space, though the term appears in several presidentially approved policy statements. The closest a president may have come to doing so was in Kennedy's May 25, 1961, speech announcing the acceleration of the program that included Project Apollo. Kennedy said that it was "time for this nation to take a clearly leading role in space achievement."[31] But as has been suggested above, Kennedy in 1962–63 wavered in his commitment to Apollo and to across-the-board preeminence in space achievements. In November 1962, Webb told Kennedy that "the objective of our national space program is to become pre-eminent in all important aspects of this endeavor."[32] The question at the time was whether to give overriding priority to Apollo, even if that meant taking funds away from other space projects. Webb argued against such a move, and Kennedy did not do so. But that does not necessarily suggest that Kennedy accepted Webb's premise, given the president's other actions before and after November 1962.

Certainly, no president since Kennedy has given the space program the priority and budget required for "a clearly leading role" in all areas of space achievement. Yet the term "preeminence" has crept into the space policy statements of the past decade. It appeared first in the 1982 space policy statement, which set as an objective to "preserve the United States' preeminence in critical major space activities to enable continued exploitation and exploration of space."[33]

By the Reagan administration's final space policy statement in 1988, the call for preeminence had been refined: "Leadership in an increasingly competitive international environment does not require United States' preeminence in all areas and disciplines of space enterprise. It does require United States' preeminence in key areas of space activity critical to achieving our national security, scientific, technical, economic, and foreign policy goals."[34]

Taken literally, this statement insists that leadership requires preeminence and that such preeminence is an important goal of U.S. space policy. At least as it applies to the civilian space program, this comes close to being a ridiculous statement, given the realities during the 1980s of funding for space overall and particularly for the development of new capabilities. The United States may declare that "preeminence in critical areas of space activity" is its goal, but no president since Kennedy has taken the actions necessary to achieve such preeminence. U.S. presidents have been willing at various times to take steps to pursue leadership of some type, but not comprehensive preeminence.

The Future

Has the quest for leadership paid off in terms of benefits to the American nation? Is space leadership worth continued pursuit? What, indeed, is the meaning of space leadership in the 1990s, almost forty years after the United States began its civilian space program?

Sally K. Ride identified well the benefits of space leadership in her 1987 report *Leadership and America's Future in Space*. She noted that during the period when the United States "was clearly and unquestionably the leader in space exploration . . . the nation reaped all the benefits of pride, international prestige, scientific advancement, and technological progress that such leadership provides."[35]

It is interesting that Ride lists pride as the first benefit of leadership. Writing almost thirty years ago, one of the first political scientists to take a hard look at the U.S. space program, Vernon van Dyke, entitled his book *Pride and Power: The Rationale of the Space Program*. Probably both van Dyke and Ride are correct—that a primary benefit of the perception of space leadership is internal to the United States. What being a leader in space says about the country seems important to its citizens. Consider only the use of space accomplishments as national symbols—they fall perhaps only behind the flag and a bald eagle in frequency. And when the United States is not doing well in space—not being the leader—this country (at least if media attention is any indicator of public interest) is concerned and wants the causes of its problems in space to be fixed.

This emphasis on pride in the U.S. leadership position is not inconsistent with the focus here on space leadership as a source of U.S. power in the world. Pride in one's country is an essential base of national power. Nations can seldom exercise influence over others when they are beset by self-doubt or lack of confidence.

Certainly, the United States "leads" in most space capabilities related to the projection of military power around the world, for example, observation and warning systems, communications satellites, global positioning systems, and so on. That leadership is an important element of the recognition that the United States is today the only global superpower in military terms.

But presidential attitudes toward leadership in national security space systems are not really the topic of this essay—though maintaining such leadership is probably even more important in today's world than it has ever been. The issue here is whether leadership in civilian space capabilities and achievements has been a worthwhile goal for this country.

Have U.S. achievements in space over the past four decades added measurably (taking that word literally) to U.S. influence in the world? In the final analysis, it is difficult to think of a way to identify and measure the *independent* contribution to U.S. international space-derived prestige of being perceived as a leader in space. There is no equation linking prestige with influence, power, and control over events and choices.

Between Kennedy's announcement of the Apollo goal and Neil Armstrong's first step onto the surface of the Moon, the United States had gotten involved in a land war in Southeast Asia, had endured a series of tense urban riots, and had suffered through a series of political assassinations. How does one calculate the difference in U.S. national prestige overall if Project Apollo had *not* been attempted, compared to that resulting from the Apollo success, in the context of everything else that shaped global attitudes toward the United States in the 1960s?

What can be said with some degree of confidence is that the notion that prestige is an important aspect of national power is well established among both students and practitioners of international relations and foreign policy. The desire to enhance national prestige has been a consistent and strong motivation of national leaders and statesmen for a long time. Reputation *is* important in dealings among nations, just as it is in dealings among individuals or firms. As this essay has pointed out, several U.S. presidents have stressed the space program as a means of enhancing the U.S. reputation in the world.

Certainly, those in the space community—and it may be possible to extend this generalization to many if not most in the United States—have internalized the expectation that the country has been, and will continue to be, a leader in space. That expectation is clearly expressed in statements of national space policy. Perhaps the primary benefit of space leadership can be found within each of us—we fully expect continued achievement in space to be part of our future. When push comes to shove, we are willing to *contribute* (today's euphemism) our tax dollars to support a strong U.S. space program.

In a very loose sense, then, the issue of U.S. space leadership—and particularly whether it continues to be worth pursuing—can be linked to the numerous analyses of America's relative standing in the world which have been so popular recently. Is America in decline, as people such as Paul

Kennedy have suggested, with the quest for space leadership part of our "imperial overstretch"?[36] Or is the United States *Bound to Lead,* with space leadership one of the bases for this country being able, as Joseph Nye suggests it must, to *manage* "the geopolitical balance of power inherited from the past, as well as the emerging interdependence that will emerge in the future."[37] This is certainly an issue for future presidents to consider as they plot the country's path into the twenty-first century.

To other sovereign nations, which are both our partners and our competitors in space and elsewhere, accepting the notion of the United States managing key areas of global relations is not particularly palatable. However, the foreign policy community around the world has been much more comfortable with the notion of U.S. leadership than the space community seems to be. Nonetheless, U.S. power, responsibly exercised, remains a key to a peaceful world with an increasingly better quality of life. What the United States does in space can—and should—be closely linked to the emerging bases of global power. As a panel of the Bush administration's Vice President's Space Policy Advisory Board observed in its January 1993 report, "As part of the United States' continuing post Cold War leadership, space achievements must be widely viewed as a key to an improved world future."[38]

This is the concept of leadership that may guide future U.S. efforts in space—taking the lead in using space to improve life on Earth. This is a different concept of space leadership than has guided most past presidents, but it is one that will likely dominate the future.

Notes

1. *National Aeronautics and Space Act of 1958,* Public Law 85–568, July 29, 1958, in *Exploring the Unknown: Selected Documents in the History of the U.S. Civil Space Program,* vol. 1: *Organizing for Exploration,* ed. John M. Logsdon, with Linda J. Lear, Jannelle Warren-Findley, Ray A. Williamson, and Dwayne A. Day (Washington, D.C.: NASA [SP-4407], 1995), pp. 334–44.

2. National Security Council, NSC 5520, "Draft Statement of Policy on U.S. Scientific Satellite Program," May 20, 1955, in *Exploring the Unknown,* ed. Logsdon et al., pp. 308–13.

3. Vernon van Dyke, *Pride and Power: The Rationale of the Space Program* (Urbana, Ill.: University of Illinois Press, 1964), p. 22.

4. See John M. Logsdon, *The Decision to Go to the Moon: Project Apollo and the National Interest* (Cambridge, Mass.: MIT Press, 1970).

5. National Security Council, NSC 5814, "U.S. Policy on Outer Space," June 20, 1958, and NSC 5814/1, "Preliminary U.S. Policy on Outer Space," Aug. 18, 1958, in *Exploring the Unknown,* ed. Logsdon et al., pp. 346, 361.

6. National Aeronautics and Space Council, "U.S. Policy on Outer Space," Jan. 26, 1960, in *Exploring the Unknown,* ed. Logsdon et al., pp. 362–73.

7. Walter A. McDougall, . . . *The Heavens and the Earth: A Political History of the Space Age* (New York: Basic Books, 1985), p. 322.

8. John F. Kennedy, "Memorandum for the Vice President," Apr. 20, 1961, in *Exploring the Unknown,* ed. Logsdon et al., p. 423.

9. Webb and McNamara to the Vice President, May 8, 1961, with the attached: "Recommendations for Our National Space Program: Changes, Policies, Goals," in *Exploring the Unknown,* ed. Logsdon et al., pp. 439–52.

10. Kennedy, "Address before the 18th General Assembly of the United Nations," Sept. 20, 1963, *Public Papers of the Presidents of the United States: John F. Kennedy, 1963* (Washington, D.C.: Government Printing Office, 1964), p. 695.

11. Kennedy to Webb, National Security Action Memo No. 271, "Cooperation with the USSR on Outer Space Matters," Nov. 12, 1963, NSF box 342, John F. Kennedy Library, Boston, Mass.

12. Kennedy, "Remarks in San Antonio at the Dedication of the Aerospace Medical Health Center," Nov. 21, 1963, *Public Papers of the Presidents of the United States: John F. Kennedy, 1963,* p. 883.

13. Johnson to Kennedy, May 13, 1963, in *Exploring the Unknown,* ed. Logsdon et al., pp. 468–73.

14. Johnson, interview by Walter Cronkite, July 5, 1969, LBJ Files, NASA Historical Reference Collection, NASA History Office, Washington, D.C.

15. Charles Towns et al., "Report of the Task Force on Space," Jan. 8, 1969, in *Exploring the Unknown,* ed. Logsdon et al., pp. 499–512.

16. Nixon, "Statement about the Future of the United States Space Program," Mar. 7, 1970, *Public Papers of the Presidents of the United States: Richard Nixon, 1970* (Washington, D.C.: Government Printing Office, 1971), p. 250.

17. Peter M. Flanigan, "Memorandum for the President," Dec. 6, 1969, in *Exploring the Unknown,* ed. Logsdon et al., p. 546.

18. Weinberger via George Shultz, memorandum for the president, "Future of NASA," Aug. 12, 1971, in *Exploring the Unknown,* ed. Logsdon et al., pp. 546–57.

19. Ibid.

20. Low, memorandum for the record, "Meeting with the President on January 5, 1972," in *Exploring the Unknown,* ed. Logsdon et al., pp. 558–59.

21. Nick MacNeil, Carter-Mondale Transition Planning Group, to Stuart Eizenstat, Al Stern, David Rubenstein, Barry Blechman, and Dick Steadman, "NASA Recommendations," Jan. 31, 1977, in *Exploring the Unknown,* ed. Logsdon et al., pp. 559–74.

22. Zbigniew Brzezinski, Presidential Directive/NSC-42, "Civil and Further National Space Policy," Oct. 10, 1978, in *Exploring the Unknown,* ed. Logsdon et al., pp. 575–78.

23. Ibid.

24. Carter, "Apollo 11 Anniversary, Remarks at a Ceremony in Observance of the 10th Anniversary of the Moon Landing," July 10, 1979, *Public Papers of the Presidents of the United States: Jimmy Carter, 1979* (Washington, D.C.: Government Printing Office, 1980), pp. 1276–77.

25. Reagan, "Radio Address to the Nation on the Space Program," Jan. 28, 1984, *Public Papers of the Presidents of the United States: Ronald Reagan, 1984* (Washington, D.C.: Government Printing Office, 1985), p. 109.

26. George M. Low, Team Leader, NASA Transition Team, to Richard Fairbanks, Director, Transition Resources and Development Group, Dec. 19, 1980, with attached: "Report of the Transition Team, National Aeronautics and Space Administration," in *Exploring the Unknown*, ed. Logsdon et al., pp. 579–86.

27. National Security Council, National Security Decision Directive No. 42, "National Space Policy," July 4, 1982, *Exploring the Unknown*, ed. Logsdon et al., pp. 590–93.

28. White House, Office of the Press Secretary, "Fact Sheet: National Space Strategy," 1984, NASA Historical Reference Collection.

29. White House, Office of the Press Secretary, "Fact Sheet: Presidential Directive on National Space Policy," Feb. 11, 1988, NASA Historical Reference Collection.

30. NASA, "Revised Talking Points for the Space Station Presentation to the President and the Cabinet Council," Nov. 30, 1983, with attached: "Presentation on the Space Station," Dec. 1, 1983, in *Exploring the Unknown*, ed. Logsdon et al., pp. 595–600.

31. Kennedy, "Presidential Address on Urgent National Needs," May 25, 1961, *Public Papers of the Presidents of the United States: John F. Kennedy, 1961* (Washington, D.C.: Government Printing Office, 1962), pp. 401–5.

32. Webb to Kennedy, Nov. 30, 1962, in *Exploring the Unknown*, ed. Logsdon et al., pp. 461–67.

33. National Security Council, "National Space Policy."

34. White House, Office of the Press Secretary, "Fact Sheet: Presidential Directive."

35. Sally K. Ride, *Leadership and America's Future in Space* (a report to the administrator), Aug. 1987, p. 5, copy in NASA Historical Reference Collection.

36. Paul M. Kennedy, *The Rise and Fall of the Great Powers: Economic Change and Military Conflict from 1500–2000* (New York: Random House, 1987).

37. Joseph S. Nye Jr., *Bound to Lead: The Changing Nature of American Power* (New York: Basic Books, 1990), p. 242.

38. Vice President's Space Policy Advisory Board, "A Post Cold War Assessment of U.S. Space Policy," Dec. 1992, p. 14 (Washington, D.C.: The White House, 1993).

Epilogue: Beyond NASA Exceptionalism

Roger D. Launius and Howard E. McCurdy

When President John F. Kennedy unveiled the commitment to undertake Project Apollo on May 25, 1961, in a speech before a joint session of Congress, he created more than a national space program to send humans to the Moon. He also created an image of presidential leadership that persisted long after the actual voyages ended. That image accentuated the power of the president to mobilize the U.S. society by setting specific national goals.

The image subsequently took on mythical qualities within the space policy subsystem. To people associated with the U.S. space program, Kennedy's mandate stood as the most significant demonstration of the conditions necessary to prosecute a successful space effort. It suggested to policy participants that extraterrestrial activities were above politics. In the years that followed, moreover, it provided an explanation for the ills afflicting the National Aeronautics and Space Administration (NASA). Those ills were frequently traced to the lack of clear mandates or presidential support for new endeavors.

People outside the space subsystem also embraced the image created by the young president, doing so to promote other agendas. Calling for presidential commitments to conquer a variety of ills, from poverty to AIDs, advocates of other initiatives advanced the notion that America could be mobilized to accomplish noble goals if only chief executives would create great mandates. Kennedy's speech stands as one of the high water marks of presidential leadership in the twentieth century, occurring as it did before images of Vietnam and Watergate tarnished the notion of executive deference.[1]

A careful examination of the record of presidential leadership in space exploration reveals that actual events differed considerably from perceptions. Space policy is not above politics. Presidential mandates do not guarantee program success. Chief executives cannot protect the civilian space agency from the forces that batter other discretionary spending programs. Space policy exceptionalism, as attractive as that notion continues to be, is not an appropriate view of reality.

If presidential leadership does not count for much, what does? After exploring the limits of presidential leadership, this epilogue examines the power of partisanship, ideology, and "pork-barrel" politics in shaping the outcome of space policy disputes involving the Congress.

The Dilemma of the "Golden Age"

Billing his speech as a second State of the Union message, Kennedy told Congress that the United States faced "an extraordinary challenge": "Our strength as well as our convictions have imposed upon this nation the role of leader in freedom's cause. . . . This nation is engaged in a long and exacting test of the future of freedom—a test which may well continue for decades to come." Kennedy argued that the United States must compete in space "if we are to win the battle that is going on around the world between freedom and tyranny." Then he added: "I believe this Nation should commit itself to achieving the goal, before this decade is out, of landing a man on the moon and returning him safely to earth. No single space project in this period will be more impressive to mankind, or more important for the long-range exploration of space; and none will be so difficult or expensive to accomplish."[2]

In announcing the Apollo commitment the president carefully gauged the mood of the nation. His rhetoric sought to capture the American imagination and overcome residual concerns regarding the difficulty and expense of the undertaking. The United States was locked in a struggle with the Soviet Union to convince uncommitted nations of the world of the superiority of the U.S. system at a time when the outcome of the cold war was far from clear. The Soviet system had bested the United States in putting both the first satellite and the first human in orbit. As John M. Logsdon has noted, "by entering the race with such a visible and dramatic commitment, the United States effectively undercut Soviet space spectaculars without doing much except announcing its intention to join the contest."[3]

A unique confluence of national anxiety, presidential commitment, scientific talent, economic prosperity, and public trust made possible the ready acceptance of the 1961 decision to carry out an aggressive lunar landing effort. Because of the president's challenge, framed as part of the cold war contest between communist nations and the "free world," NASA undertook a mobilization comparable, in relative scale, to the Manhattan Project in World War II and the national program to deploy an intercontinental ballistic missile (ICBM) system during the 1950s. Accordingly, the space agency's annual budget increased from $500 million in 1960 to $5.2 billion in 1965, as government and industry executives mobilized to combat what they considered a national crisis. Project Apollo, backed by sufficient funding, provided a tangible result of the power that a national commitment in response to a perceived threat to U.S. welfare could have.[4]

As part of his speech on "Urgent National Needs," Kennedy announced twelve specific legislative proposals, of which only four dealt with space. The disposition of the other eight items in Kennedy's speech contrasted sharply with the congressional response to his rhetoric on space. Of the other eight initiatives, Congress approved only four. It failed to enact the president's Manpower Development and Training Program, refused to provide him with a foreign aid contingency fund, and voted down his request for a nationwide system of fallout shelters to protect the populace "in the event of a large-scale nuclear attack." The fate of those eight items mirrored Kennedy's overall legislative success rate in 1961. During his first year in office, Kennedy submitted 355 specific requests to the Congress. He won approval for 172, or slightly less than half. In historical perspective, that was a fairly low box score for a modern president in his first year of office.[5]

Kennedy worried that the Congress might treat his space initiatives with similar disrespect. Congressional debate was perfunctory, however, and NASA found itself pressing to spend the funds committed to it during the early 1960s. Congress approved all four space initiatives, including the first steps toward the Moon, as Logsdon has noted, "almost without a murmur."[6] Kennedy's speech subsequently took on mythical qualities as a demonstration of the president's power to set the national space agenda and stir the American imagination. With the speech and the events that preceded it, the president established a national objective, galvanized public support, enlisted the Congress, and set in motion the activities that mobilized a nation. The symbolism offered by Kennedy's decision was applied subsequently to a wide range of issues, from poverty and housing to health care and the environment ("if we can send a man to the Moon, why can't we clean up Chesapeake Bay?").[7]

The symbolism has held special appeal for space boosters, the community of true believers who promote space exploration as one of the nation's most enduring legacies. To them, the lunar decision suggested that space exploration deserved special treatment within the American political system. Like foreign policy, space exploration seemed to enjoy a higher priority than domestic endeavors. Space exploration seemed to be above politics, meaning that it received far-reaching bipartisan support. Congress and the public seemed to defer to the president, generally acquiescing in his choice of objectives. The decision to go to the Moon suggested that a president could overcome partisan divisions and lead the nation to great accomplishments, if only the objective was properly framed. Indeed, many have argued that the subsequent ills of the space program can be traced to the unwillingness of more recent presidents to make "Apollo-like" public commitments.[8]

Project Apollo, while an enormous achievement, left a divided legacy for space enthusiasts. The sprint to the Moon transformed the national civil space organization into an agency preoccupied with a single mission. Single-issue agencies—like single-issue political parties—have a difficult time dealing with success. A search for continued meaning inevitably ensues. Moreover, the results of Apollo, in contrast to what space buffs had wanted, largely proved to be a technological deadend. As the sometime senior NASA official Hans Mark observed, "President Kennedy's objective was duly accomplished, but we paid a price: the Apollo program had no logical legacy."[9]

The "golden age" of Apollo also created an expectation that the issuance of a major space goal by the president would always bring to NASA a broad consensus of support and provide it with resources as well as the license to dispense them as agency leaders saw fit. Most space advocates did not understand at the time how exceptional the Apollo mandate was. The Apollo decision was an anomaly in the history of the U.S. space program.[10] Officials within NASA came to perceive as normal the free-flowing funds, ready political support, and relative autonomy when such was rarely the case.[11] This was the dilemma of the golden age of space exploration. Program success was predicated upon a set of conditions that could hardly be restored.

The Limits of Presidential Leadership

History shows that presidents prevail on more space issues than they lose. Yet, while presidents win on most of the space initiatives they propose, they are by no means omnipotent. They face significant opposition to initiatives that succeed, and they occasionally fail. Their success rate on space initiatives is not much different from the success rate for presidential initiatives in general. In its annual review of congressional roll call votes, the Congressional Quarterly research service reports that presidents prevail on about 70 percent of the key votes on which they take a position.[12] The success rate varies widely from year to year. Roll call votes do not cover all initiatives proposed (some space policy questions are decided in committee), but the box scores suggest that the perception of presidential success on space issues may be due in part to confusion about the extent of presidential success on issues of all kinds.

Presidents certainly have prevailed on a succession of space policy issues. They have not done so, however, without having to deal with opposition. President Dwight D. Eisenhower, the first chief executive in the space age, played a central role in the decision to create the civilian space program. In the face of military demands for control of the program, he used a presidential directive to assign NASA the responsibility for Project Mercury (the program to put the first astronauts in space).[13] Once underway, however, the sacrosanct Project Mercury garnered its share of congressional discord. Edward Boland, then a fourth-term congressman from Massachusetts, objected to a Republican effort to add extra funds to the NASA research and development budget on the grounds that the newly created space agency had already received more money than it could wisely spend. The House voted to support the president's position by a margin of two to one. Congressional challenges subsequently emerged to presidential initiatives in Project Apollo, the space shuttle program, and various iterations of the NASA space station.

Not all presidential victories cast the chief executive in the advantageous position of advocating new ideas while placing opponents in the stingy position of saying "no." In 1970, President Richard M. Nixon asked Congress to cut NASA's budget to $3.4 billion, a significant reduction from the $5 billion budgets that NASA had enjoyed during the mid-1960s. Space boosters pulled a classic end run, convincing members of the House to authorize a $300 million increase in the human spaceflight budget.[14] After a struggle, the president prevailed, as Senators and conferees removed most of the increase.

During the 1970s, presidents proposed and won approval for a series of large science projects: the Hubble Space Telescope, the *Pioneer* probe to Venus, and the *Galileo* probe to Jupiter. All three programs faced strong opposition centered in the House Appropriations Committee handling NASA affairs. In 1974, following the recommendation of Boland, the

appropriations subcommittee chair, the full House of Representatives deleted funds for the Hubble Space Telescope. After the Senate restored telescope funds, the House in 1975 deleted most of the funds for the *Pioneer*
orbiter and lander on the grounds that NASA should not undertake two
big science projects at one time. Again, the Senate restored the funds. In
1977, the House axed funds to start the *Galileo* probe during initial consideration of the NASA appropriation bill. Following conference committee
negotiations with the Senate, Boland asked for and lost a vote of support
from the whole House. In a stunning reversal for Boland, the House voted 131 to 280 to support the president's position in favor of the probe.[15]

The notion that presidents can avert opposition simply by making a
strong public statement committing the nation to a specific "big-ticket"
goal is not supported by the historical record. Not even the rhetoric of
Kennedy's Apollo decision stood up to the pressures of later years. The
events of 1963 bear this out. Two years after Kennedy challenged the nation to race to the Moon, he undercut his rationale for the adventure by
proposing in a September 1963 speech at the United Nations that the Soviet Union join the United States in completing the voyage. Representative Thomas Pelly (R-WA) stood up in the well of the House three weeks
later and offered an amendment to prohibit the use of government funds
to finance a joint expedition. In spite of Kennedy's insistence that his U.N.
proposal merely carried out the mandate for international cooperation in
NASA's enabling legislation, Pelly's amendment passed.[16]

Buoyed by the thought that the United States and Soviet Union were
no longer racing to the Moon, opponents of Project Apollo moved to cut
funds for what one called "a manned junket to the moon." The House
began the assault by removing $600 million from Kennedy's $5.7 billion
NASA budget request. The administration appealed to the Senate, arguing that nothing less than $5.4 billion would keep Project Apollo on sched-
ule. Rather than receiving sympathy, the Kennedy administration faced new
levels of hostility. Arkansas Senator J. William Fulbright moved to cut 10
percent more from the NASA appropriation. The president's allies prevailed
on that vote but failed to sustain a Senate Appropriations Committee recommendation to add $90 million to the House figure when Wisconsin
Senator William Proxmire successfully moved to strike that amount.
Proxmire's victory on a 40 to 39 vote clearly marked an end to Kennedy's
ability to charm the Congress with his visions of space exploration.[17]
Kennedy's 1961 speech initiating Project Apollo may have been a special
moment, but it by no means created a form of magic whose rituals guar-
anteed future success.

Since then, as Logsdon points out in the chapter 7 of this book, the president has continued to be the primary determinant "of the content and pace of the civilian space program." The president, however, wields that power within a sophisticated structure of opportunities and constraints. Many factors tug at the loyalties of people outside of the White House, of which loyalty to the president is only one. The 1991–93 debate over the future of the NASA space station reveals precisely how much of a difference presidential leadership can make.

In 1991, after begrudgingly approving six years of development funds for the space station, the House Appropriations Subcommittee handling NASA's budget voted to kill the controversial project. Administration officials appealed the decision to the whole House. They lobbied hard to have the larger assembly reverse the committee recommendation, always a difficult task in a legislative body given to committee deference. President George Bush and Vice President Dan Quayle personally joined the effort to gain funding for continued space station development, contacting members from key districts. Heavy lobbying produced a 240 to 173 space station resurrection on the House floor. The issue resurfaced again in 1992. This time the appropriations subcommittee chair, Bob Traxler of Michigan, forced an up-or-down vote on the space station on the House floor. Another round of intense lobbying followed, and the administration position prevailed by a vote of 237 to 181.[18]

NASA officials believed that the issue was behind them, but redistricting and the 1992 election produced a large class of freshmen legislators who had never voted for the controversial project. Skepticism about the cost and value of the space station enlarged the opposition and threatened the project anew. Again, the president supported the project. It was a scaled-down project and a new president (Bill Clinton), but it had presidential support nonetheless. Most of the old members of the House who had voted on the issue through the two previous challenges did not change their positions, but some did. An analysis of their behavior provides insights into the extent of presidential influence.

The key vote came on an amendment to eliminate a seven-year $12.7 billion authorization and terminate the project. Twelve Republicans who had voted for the space station under Bush voted "nay" under Clinton. It may be surmised that they never possessed a great deal of enthusiasm for the project but initially voted for it out of deference to Bush. Once he was gone, however, their support disappeared.

Thirteen Democrats changed their votes in the opposite direction. They had opposed the space station in 1991 and 1992, but with their own president in the White House, they supported the project. The most dramatic switch was provided by Louis Stokes (D-OH), the new chair of the House Appropriations Subcommittee handling the bill. In 1991 and 1992, as a majority member on the subcommittee, he had voted to kill the space station. Motivated in part by a desire to please Clinton, Stokes supported the space station in 1993.

Twelve votes lost and thirteen found produced a net gain of one vote for presidential leadership. That is not much in a body with 435 members (plus five delegates who can vote on floor amendments). The shifts nearly canceled out each other. Given the fact that the space station survived the 1993 challenge by a single vote (215 to 216, including the delegates), small shifts provided space station supporters with all they needed to prevail. In that sense, presidential leadership made an important contribution.[19]

Taken as a whole, however, the history of civilian space policy provides little support for the doctrine of presidential omnipotence. While presidential leadership makes a difference, that difference is marginal. Presidents prevail most, though not all of the time. Presidential success in space policy is not remarkably different from presidential success with Congress on all policies. The notion that presidential leadership in space somehow provides the chief executive with special opportunities for success is not supported by the whole record.

The President and Space Policy Losses

While presidents have played the dominant role in setting the national space agenda, leading the nation toward that agenda has been a constant struggle. The limits of presidential leadership in the U.S. space program and the factors that shape the president's ability to lead have been very real, though often forgotten in the glare of the successful race to the Moon. The Apollo episode suggested to many people associated with the space program that once a president committed the nation to a specific objective and placed the power of the Executive Office behind it, the initiative would move forward. This belief gained added credibility as subsequent presidents were able to overcome early opposition to the next two human spaceflight initiatives—the space shuttle program and the *Freedom* space station.[20]

This view is not historically valid. The most dramatic exception to the record of presidential preeminence occurred in July 1989 when President Bush proposed an ambitious Space Exploration Initiative (SEI) that would return Americans to the Moon, establish a lunar base, and then, beginning from a NASA-built space station, send human expeditions to the planet

Mars. In advancing SEI, Bush followed the classic script for exercising leadership in space. He made a Kennedy-like announcement, complete with a strong personal commitment, proposing the initiative during a major address commemorating the twentieth anniversary of the first landing on the Moon delivered from the steps of the National Air and Space Museum with the *Apollo 11* astronauts at his side. Like Kennedy some thirty years earlier, Bush subsequently elaborated on his proposal in a speech at a Texas university (Kennedy spoke at Rice University; Bush chose Texas A&M). Having set a specific objective, with added deadlines, Bush instructed his National Space Council to marshal the power of the Executive Office behind the proposal.

While Bush had announced, à la Kennedy's Apollo decision, a dramatic space exploration project, the similarities between those two presidential initiatives began and ended with the public declarations. Whereas the 1961 Apollo decision received broad national support, continually nurtured by senior officials at NASA and other sectors of the government, support for the Bush program was tenuous at best and could not be maintained during the years that followed.[21]

Public and congressional reaction to the Space Exploration Initiative was lukewarm from the start, especially when budget estimates for the overall endeavor began flowing in. Representative Leon Panetta criticized Bush for "talking promises . . . without any regard to the fiscal consequences."[22] The *Washington Post* observed that "easy slogans are not going to conquer the high hurdles of politics. This is not 20 years ago, when the ringing words of a president could spark a commitment of some 4 percent of the U.S. budget toward a trip to the moon by a certain date."[23]

Representative Bob Traxler and Senator Barbara Mikulski, chairs of the House and Senate Appropriations Subcommittees that acted on the NASA budget, immediately cut most of the funds that Bush had requested for starting the initiative. Normally strong supporters of the civilian space program, they directed NASA to focus its attention on projects already underway. In the following year, funding for SEI was virtually zeroed out of the budget despite lobbying from Vice President Quayle, head of the president's National Space Council. Although Bush castigated Congress for not "investing in America's future," members decided to spend the funds elsewhere. "We're essentially not doing Moon-Mars," Mikulski bluntly declared.[24] As a result, the Space Exploration Initiative died a quiet death on Capitol Hill.

Rejection of SEI provides a spectacular illustration of the limits of presidential leadership in space. It is not the only example. During the period when NASA struggled to complete Project Apollo, Congress altered or terminated three major presidential space initiatives. The first of these was Project Voyager, a presidentially supported mission to Mars. Based on a recommendation from the National Academy of Sciences, NASA officials formulated and won presidential approval for $2 billion to launch a pair of robotic orbiter-landers on one Saturn V rocket on a mission to explore the Red Planet. Even with the endorsement of the Johnson administration, Voyager was controversial from the start. A few scientists supported the mission but many opposed it as technically ambitious, exceptionally risky, and overly expensive. In the summer of 1967, following conflicting testimony from scientists and with a shortage of funds for other priorities like the Vietnam War, Congress denied the money necessary to begin the project.[25]

NASA leaders and their scientific clientele learned, at least temporarily, several hard lessons from the Voyager failure. Most importantly, they realized that strife within the scientific community had to be made invisible in order to put forward a united front against the priorities of other interest groups and other government leaders. While scientific consensus could not guarantee that any initiative would become political reality, without it a program could not be funded. Supporters also learned to respect the limits of pricing. While a $750 million program might encounter little political opposition, a $2 billion project crossed an ill-defined but very real threshold triggering intense competition for those funds. Finally, supporters learned that presidential sympathy for a "big science" project was no guarantee of congressional support. Having learned these lessons, as well as some more subtle ones, the space science community regrouped and returned with a trimmed-down Mars lander program, Project Viking, that was funded and provided good scientific return in the mid-1970s.[26]

Also during 1967 Congress sharply reduced funds for the Apollo Applications Program, a series of undertakings designed to keep the United States in space after the flights to the Moon. President Johnson, under pressure to fund the Vietnam War as well as a variety of Great Society programs, submitted a scaled back NASA budget that contained $455 million for Apollo Applications. At the heart of this initiative was a modest orbital workshop, later named *Skylab,* that could be flown from the ground and occupied occasionally by astronauts. It would be, NASA officials hoped, the precursor of a real space station. Congress appropriated only $348 million, essentially dismantling the program schedule. NASA later realized some success with the program in 1973 as three human spaceflight missions were flown to one orbiting facility.[27]

Finally, in 1971 Congress cut President Nixon's request to fund a series of spacecraft that would undertake a "Grand Tour" of the outer planets of the solar system. Nixon's 1971 budget request included $30 million to initiate this $1 billion project. Congress appropriated only enough money to allow NASA to study other alternatives.[28] Over the objections of NASA Administrator James C. Fletcher, Nixon's staff resubmitted a much slimmed-down program the following year as a two-planet mission to Jupiter and Saturn called Project Voyager. At the time, this was viewed as a major political defeat for space scientists, overcome only when NASA technology produced a spacecraft that far exceeded its design specifications and flew on to Uranus and Neptune.[29]

Partisanship and Space

The fact that presidents sometimes fail, or prevail only after a struggle, indicates the presence of other forces beyond presidential leadership that work to affect the overall level of support for space policy. Partisanship, ideology, and "pork-barrel" politics also play a significant role.

For much of its history, space has not been a deeply partisan issue. Once in the White House, both Democrat and Republican presidents have become advocates for space initiatives. Both Democrat and Republican presidents have chosen to restrain them. In Congress, space boosters have relied upon allies within both political parties to fight for new initiatives and rescue old undertakings. Opponents as well can be found on both sides of the aisle. The presence of bipartisan support should not be taken as a sign that space policy is somehow above politics, however, or that it enjoys special immunity from the level of partisanship that affects other endeavors. In fact, space policy tends to mirror the ebb and flow of partisan divisions within the political system at large.

When the space age began with the launching of *Sputnik I* in 1957, partisan differences emerged. Democrats in Congress seized upon the space issue, as David Callahan and Fred I. Greenstein point out in chapter 1 of this book, to underscore "a broader failure by Eisenhower and other Republican leaders to provide sound national leadership." Democratic leaders complained about Eisenhower's "beginner" space program and threatened direct action if the leaders of the newly created NASA did not show "proper imagination and drive."[30] Democrats questioned Eisenhower's executive abilities and put forth an image of the president as a smiling incompetent. G. Mennen Williams, the Democratic governor of Michigan, even released a poem linking Eisenhower's inaction to his well-known fondness for golf.

Oh little Sputnik, flying high
With made-in-Moscow beep,
You tell the world it's a Commie sky
and Uncle Sam's asleep.
You say on fairway and on rough
The Kremlin knows it all,
We hope our golfer knows enough
To get us on the ball.[31]

Democratic Senator and presidential contender Lyndon B. Johnson opened hearings before a subcommittee of the Senate Armed Services Committee beginning on November 25, 1957, to assess the impact of *Sputnik I*, embarrassing the president and the Republican party. In 1960, John F. Kennedy ran for the presidency with Johnson as his running mate in a campaign that labeled the Republican Eisenhower as a "do-nothing" president. Kennedy was especially hard on Eisenhower's record in international affairs, exposing the so-called "missile gap" with the Soviet Union as an example of Eisenhower's inattentiveness (a charge that later turned out to be false).[32]

Republicans returned fire on the Kennedy administration during the early 1960s. Ex-president Eisenhower questioned the wisdom of spending more than $20 billion on what he called "a mad effort to win a stunt race" to the Moon: "Why the great hurry to get to the moon and the planets? We have already demonstrated that in everything except the power of our booster rockets we are leading the world in scientific space exploration. From here on, I think we should proceed in a orderly, scientific way, building one accomplishment on another."[33] In a later article entitled "Why I Am a Republican," Eisenhower argued that the Moon race "has diverted a disproportionate share of our brain-power and research facilities from equally significant problems, including education and automation."[34]

Eisenhower's NASA administrator, T. Keith Glennan, provided substance for the party line. Corresponding with members of the Eisenhower administration, Glennan expressed misgivings about the commitment to race the Soviets to the Moon.[35] He told Eisenhower, then in retirement at Gettysburg, Pennsylvania, "This is a very bad move. . . . We are entering into a competition which will be exceedingly costly and which will take up an increasingly large share of that small portion of the nation's budget which might be called controllable."[36]

Responding to this party line, Republicans forced Democratic lawmakers to defend Project Apollo against other priorities. In 1963, Representative Louis Wyman (R-NH) moved to cut $200 million from the NASA research and development budget. Wyman argued that the United States could not afford a crash program to reach the Moon when defense needs pressed harder. His motion was defeated in a rare party line vote on space, with 90 percent of the Democrats supporting Project Apollo and 89 percent of the Republicans supporting the cut.[37]

Four years later, the space program faced a similar challenge. Senator Proxmire moved to cut $361 million from NASA's overall authorization. This time a bipartisan coalition of Republicans and Democrats joined together to turn back the assault. Fifty-eight percent of the Democrats and 61 percent of the Republicans present voted to deny Proxmire's request.[38] This was followed by roll call votes during the 1970s in which a similar bipartisan coalition of Democrats and Republicans voted to start up projects like the space shuttle and *Galileo*. The votes suggested that space policy had reached a level of political maturity wherein it enjoyed support from both political parties. This notion is somewhat disingenuous.

While it is true that space exploration enjoyed bipartisan support during that period, it would be premature to conclude from instances such as these that space policy enjoyed a status above politics. The decline in party line votes over space policy to a certain degree reflected the decline in party line votes in general. Between 1961 and 1976 the proportion of party line votes (in which a majority of Democrats opposed a majority of Republicans) fell from 50 to 36 percent in the House and from 62 to 37 percent in the Senate.[39] This continued a trend that had been underway for nearly one hundred years. Partisanship on Capitol Hill had been in gradual free fall since the late 1800s, when roughly three out of four votes in the House pitted a majority of one party against the other.[40] Given the declining importance of partisanship in the national government overall, it should not be surprising to find so few party line votes over space policy in the late 1960s and 1970s.

The tradition of bipartisanship continued into the 1980s. When President Ronald Reagan proposed in 1984 that the United States start work on a permanently occupied space station, for example, he found support and opposition in both parties. Congressman Bill Green (R-NY) helped lead opposition to the project on the House Appropriations Subcommittee handling the money bill. Senator Jake Garn (R-UT), who flew on the space shuttle in 1985, gave the White House fits with his calls for automation on what NASA had promised would be a fully human facility. Space station supporters had to rely upon a coalition of junior members (both Democrats and Republicans) to move the money bill out of the House Appropriations Subcommittee where Green and Boland, the subcommittee chair, would have happily disapproved it.

As the space station program encountered further difficulties, partisan differences reappeared. The project, troubled by growing costs and missed deadlines, became more closely associated with the Republican administration that had nurtured it through various predicaments. It consequently lost support among House and Senate Democrats. In 1988, Republican lawmakers mobilized to defeat an effort to remove $400 million from the NASA budget. House Republicans mobilized 77 percent of their members to defeat the motion. The Senate followed in 1991. A move by Dale Bumpers (D-AR) to gut the space station program won a ten-vote majority (32 to 24) among Senate Democrats. Republican Senators supported the space station by a margin of 40 to 3 in order to maintain the initiative.

By the late 1980s, partisanship had reemerged in the politics of space exploration. Seven of the eight most important votes that civilian space boosters faced on Capitol Hill in 1988–93 found a majority of Democrats opposing a majority of Republicans. In 1993, the space station faced two challenges in the House and one in the Senate. Republicans found themselves in the unusual position of supporting Democratic President Clinton, whose position in favor of the project was opposed by three-fifths of the members of his own party.

The revival of partisan differences on space policy followed a larger trend toward partisanship over all manner of issues. All told, the number of roll call votes in which party opposed party (a majority of each) rose from 36 percent in 1976 to 65 percent in 1993 on the floor of the House. Senate partisanship increased in a similar manner.[41] The growth in partisanship was associated with an increase in the ideological shrillness of American politics.[42]

Space policy has not enjoyed a charmed existence, immune from the residual level of partisanship affecting the national government as a whole. It has, to be sure, enjoyed a measure of bipartisan support, especially from the late 1960s through the 1970s. Partisan differences over space have moved through cycles more or less like those affecting government in general. Party affiliation is not a strong predictor of positions on space, but neither is it irrelevant. The idea that presidential leadership somehow allows the space program to sit on a plateau above partisan differences is simply not supported by the historical record.

Political Ideology and the Politics of the Space Program

If presidential leadership and bipartisan agreement do not affect space policy to a greater degree than other issues, then what does? What has motivated politicians and other leaders to either embrace or reject space exploration? Historically, two factors have shaped the space policy debate in American politics: ideology and the pragmatism of the government contract. Among all the factors affecting space exploration, ideology is the most important. Politicians come to Washington with agendas to complete. Their attitudes toward space exploration are strongly influenced by those agendas. Only when countervailing forces are strong do politicians move off of their ideological agendas. This is not to say, however, that ideology provides a stable continuum for predicting individual positions. Ideological perspectives on the value of a strong space program have undergone a complete reversal since the space age began.[43]

From the beginning of the space age in 1957, the ideological debate over the program has revolved around the expense and direction of the enterprise, particularly the emphasis placed on human spaceflight initiatives as opposed to scientific objectives. In the 1950s, conservative leaders favored a limited civilian space program based on scientific endeavors. Within the Congress, this approach was embraced by conservative southern Democrats who did not want to see the federal budget grow. The standard-bearer for this approach in the executive branch was Eisenhower, whose preoccupation with the need to conduct survelliance flights over the Soviet Union shaped much of his attitude toward space.[44]

Eisenhower supported an extensive program of reconnaissance satellites as a means of checking potentially aggressive actions by the Soviet Union. The safety from surprise attack promised by reconnaissance satellites was an especially attractive feature because Eisenhower and leaders of his generation remembered well the Japanese attack at Pearl Harbor on December 7, 1941, and were committed to never falling for such a sucker punch again. At a meeting of key scientific advisors on March 27, 1954, to discuss the use of space for military purposes, Eisenhower warned that "modern weapons had made it easier for a hostile nation with a closed society to plan an attack in secrecy."[45] Launch of an Earth-orbiting satellite during the International Geophysical Year (IGY) was viewed by some advisors within his administration as a cold war measure. By linking the first satellite launch to an international scientific effort, Eisenhower could establish the principle of overflight, namely, that a satellite orbiting over enemy territory did not constitute a hostile act. Establishing the principle of peaceful overflight was a key reason behind Eisenhower's desire to move scientific research in space out of the military and into a civilian agency.[46]

Following this logic, conservatives favored a civilian space program with clear scientific objectives whose expenses were modest enough not to distract from more important national security needs. Even the proposed "man in space" program was viewed in conservative circles not as an exploration program but as a means to test the effects of radiation and weightlessness on human beings who might be called upon to orbit the Earth.[47]

This conservative approach toward space exploration dominated U.S. policy making until the Soviet launch of *Sputnik I* on October 4, 1957. Suddenly, supporters of a restrained civilian space program awoke to intense criticism. Critics ballyhooed the illusion of a technology gap and demanded increased spending for aerospace endeavors, technical education, and new federal agencies. The call for an aggressive space program was especially prominant among liberals in the Senate, such as Johnson and Kennedy. This division led to the conclusion that ideological conservatives favored a small, diffident space program while those with liberal perspectives advocated a much more forceful and dynamic effort.[48]

Indeed, liberals pushed hard for an aggressive human spaceflight program with large federal expenditures presided over by a huge federal presence (NASA). From the outset, liberals demanded that sufficient funds be appropriated to ensure national prestige in the international arena, something that conservatives eschewed. T. Keith Glennan, NASA's first administrator, argued against entering into a "space race" with the Soviet Union in a typically conservative letter to his successor, James E. Webb.

> There can be only one real reason for such a "race." That reason must be "prestige." The present program without such a "race" but with full intention of accomplishing whatever needs to be accomplished . . . will produce most of the significant technology and essentially all of the scientific knowledge that will be produced under the impetus of the "race" and at the lower cost in men and money. . . .
>
> I don't think we should play the game according to the rules laid down by our adversary. I do believe that such prestige is apt to be less than enduring as compared to the respect and friendship we will gain from foreign aid programs, well administered over the same six or eight years.[49]

Senate liberals pressed to engage the Soviet Union in such a race, and President Kennedy's decision to undertake Project Apollo accomplished just that. In 1963, Senator Fulbright, generally classified as a conservative on fiscal matters, moved to strike $519 million from the NASA budget.[50] Republicans and southern Democrats, the groups among which most conservatives could be found, split evenly on the issue. Northern Democrats rescued the program. Indeed, the northern Democrats voting to rescue the NASA space program read like a who's who of American liberalism: Edward Kennedy of Massachusetts, Hubert Humphrey and Eugene McCarthy of Minnesota, Edmund Muskie of Maine, Warren Magnuson of Washington, and Abraham Ribicoff of Connecticut.

By 1967, ideological divisions had begun to shift. A similar motion by longtime critic Proxmire now found northern Democrats divided on the wisdom of NASA spending, with the more conservative Republicans and southern Democrats coming to its aid.[51] The conversion completed itself with Nixon's election to the White House in 1969. Shortly after his inauguration, Nixon received recommendations regarding the future of the U.S. space program beyond the flights to the Moon. Space advocates urged him to continue the spending levels attained by NASA during the Apollo years. As leader of the party that had initially opposed the race to the Moon, Nixon could have wound down the manned space program in favor of a more conservative approach.[52] Instead, following a contentious White House staff debate, Nixon and his aides decided to maintain a modest spaceflight program by approving construction of the space shuttle. In terms that recalled Kennedy's original rationale, Caspar W. Weinberger, deputy director of the president's Office of Management and Budget (OMB), explained why the White House overcame conservative inhibitions:

> This was the next frontier, and it [the space shuttle] was the one thing that would enable us to achieve a very substantial direct return from the huge investment in space. To me, it would have been all right to invest the amount necessary to go to the moon, simply because of the fact that it was a terribly important thing for America to do. . . . It seemed to me that we could either grasp it [the shuttle technology], or fall irretrievably behind, and it was something that therefore we should do.[53]

John Ehrlichman, the president's senior advisor on domestic affairs, listed a second reason for the president's approval of the program. "He liked heroes," Ehrlichman recalled. Nixon viewed the astronauts as embodiments of traditional American ideals and did not want to be the president who removed them from the national scene.

> He thought it was good for this country to have heroes. The country didn't have very many heroes. But he had a whole lot of heroes as a boy, and he was a reader of history. I think that was part of it, too. He had this kind of metaphysical thing about national morality, national fiber, and national ideals, which he would spin out; and . . . he would sit and just sort of ruminate on these things—drugs and decline of American moral fiber and all of that, [which] somehow or another had to be dealt with by the President.[54]

Conservative support for the space shuttle within the Nixon administration was further advanced by the national security applications of shuttle

technology. The Department of Defense (DOD) planned to use the shuttle to launch U.S. reconaissance and communications satellites on short
notice. Ehrlichman even thought that the shuttle might be used to capture
enemy satellites.[55] Nixon was very impressed with the shuttle's potential
for military missions and emphasized this outlook in supporting the project.
The support of Air Force Secretary Robert C. Seamans Jr. and other Defense Department officials for national security use of the shuttle helped
to make the program more palatable to conservatives.[56]

As conservative support shifted toward the space program, liberal support moved away. Minnesota Senator Walter Mondale, a leading liberal,
introduced a 1972 amendment to delete all funding for the space shuttle.
In 1973, then-Representative Ed Koch of New York complained about
space exploration in general. "I just for the life of me can't see voting for
monies to find out whether or not there is some microbe on Mars, when
in fact I know there are rats in the Harlem apartments."[57] Even liberals
who supported the space program complained about excessive spending,
a traditionally conservative concern. Representative George E. Brown Jr.
(D-CA) noted in 1992 "that some of our proudest achievements in the
space program have been accomplished within a stagnant, no growth
budget." He applauded the science programs of the 1970s "when the
NASA budget was flat. It would be wise to review how we set priorities
and managed programs during this productive time."[58]

The sea change in ideological attitudes toward space went considerably
beyond the happenstance of presidential control. It drew its strength from
the confluence of two broad forces in American political thought: the
changing nature of American liberalism and the conservative embrace of
frontier mythology.

President Kennedy had raised the analogy of the frontier as a rationale
for his ambitious vision of space exploration.[59] As the space program
matured, liberals abandoned this point of view. For Kennedy, space as a
new frontier was inextricably tied to the cold war. Once liberal interest in
the cold war waned, so did the necessity of dominating "this new sea."[60]
Moreover, liberals grew increasingly restless with the exploitation and
oppression that the frontier myth seemed to imply. Conservative thinkers
suffered no such misgivings. To them, space remained a "new frontier" that
allowed them to contemplate separation from a stagnant civilization, to
struggle against a harsh environment, and to reap the benefits of the economic progress that (in their minds) would inevitably follow.

President Nixon referred directly to these images on January 5, 1972,
when he approved the space shuttle. He stated that the United States should

build the system because it would "help transform the space frontier of the 1970s into familiar territory, easily accessible for human endeavor in the 1980s and '90s." He closed with a quote from Oliver Wendell Holmes about sailing into the unknown.[61] On January 25, 1984, President Reagan used similar references as he directed NASA to start work on a space station. "Our second great goal," he said, "is to build on America's pioneer spirit." He characterized space as "our next frontier" and called the station a means for America to rekindle the advantages of past frontiering. "Just as the oceans opened up a new world for clipper ships and Yankee traders," he promised, "space holds enormous potential for commerce today."[62]

The linkage of the space program to traditional ideas about the frontier has been an important ingredient in the overall effort to build conservative support. The popular conception of "westering" and the settlement of the American continent by Europeans from the east has been a powerful metaphor for the propriety of space exploration and has enjoyed wide usage by conservative supporters. Its images of territorial discovery, exploration, colonization, and exploitation represent positive ideals for these people, if not for others.[63]

Liberals have come to view the space program from a quite different perspective. To the extent that space represents a new frontier, it conjures up images of commercial exploitation and the subjugation of oppressed people. Implemented through a large aerospace industry, it appears to create the sort of governmental-corporate complexes of which liberals are increasingly wary. Despite the promise that the shuttle, like jet aircraft, would make spaceflight accessible to the "common man," space travel remains the province of a favored few, perpetuating inequalities rather than leveling differences. Space exploration has remained largely a male frontier, with little room for women and minorities. In the eyes of liberals, space perpetuates the inequalities that they have increasingly sought to abolish on Earth. As a consequence, it is not viewed favorably by those caught up in what the political scientist Aaron Wildavsky has characterized as "the rise of radical egalitarianism."[64] The advent of this new liberal philosophy coincided with the shift in ideological positions on the U.S. space program in the late 1960s.

Writing from this new tradition, the western historian Patricia Nelson Limerick has argued that the frontier myth should not be employed as a happy metaphor by space boosters but as a pejorative reflection. The frontier metaphor, she argues, denotes conquest of place and people, exploitation without environmental concern, wastefulness, political corruption, executive misbehavior, shoddy construction, brutal labor relations, and financial inefficiency. Limerick feigned surprise that no one from NASA had attacked those employing the frontier analogy "for insulting the organization's honor. It's a wonder no one—no shuttle pilot, mission coordinator, mechanic, or technician—said, 'Now cut that out—we may have our problems, but it's nowhere near that bad.'"[65]

The civilian space program began with conservatives embracing a limited undertaking with modest scientific objectives. Liberal Democrats created a crash program supported by a robust aerospace industry. Conservatives found it easier to vote for space spending as the program matured, especially when industrial contracts were directed toward conservative strongholds in the South and West. Liberals did not. By the 1980s, the transformation was complete. Liberals in both parties found themselves opposed to the big space program, while conservatives (increasingly concentrated in the Republican party) had become the bearers of the Kennedy appeal.

The Economic Imperative and Political Leadership in Space

Just as ideology has been important in shaping levels of support for the space exploration, so has the influence of government spending. NASA's political base is broadened by the fact that spaceflight activities are carried out at NASA field centers in ten different states. NASA contracts out nearly 90 percent of its budget, further amplifying the economic impact of its activities. As numerous space boosters have observed, not one dollar has been spent *in* space. All of the money has been spent on the ground.

Some people view space exploration as a scientific "pork barrel" from which federal funds are distributed. It is true that government contracts help to create support for the space program in areas of ideological opposition.[66] Spending on space has not been as powerful as ideology or party in recent years in resolving policy disputes, but it does play a significant role.

The location of NASA field centers has followed political necessity as well as technical requirements. Glennan reported on a 1959 meeting with Albert Thomas, the Texas Democrat who chaired NASA's Appropriations Subcommittee. Thomas informed Glennan that Rice University would provide a thousand acres of land for a NASA installation south of Houston and urged Glennan to take it. Glennan hesitated, knowing that NASA intended to merge its human spaceflight and satellite programs at the new Goddard Space Flight Center north of Washington, D.C. NASA had "other things to do, much more important than building buildings," he told Thomas. Thomas countered by explaining that Glennan would not get the money he needed for the Goddard facility unless he put one in Houston, too. NASA eventually built the Johnson Space Center in Texas, which became the home for its human spaceflight programs.[67]

Government contracts flowed out from NASA field centers both old and new as the agency matured. Rather than conduct the bulk of their work in-house, NASA officials contracted out between 80 and 90 percent of the funds they received each year. This was a deliberate strategy initiated by NASA's first two administrators, Glennan and Webb, both to build up the capabilities of the American aerospace industry and to build a broad base of political support for the infant space program. For Glennan and his ideologically sympathetic boss, Eisenhower, reliance on the private sector came naturally. Glennan wrote that he came to NASA with "a firm conviction that our governmental operations were growing too large, [and] I was determined to avoid excessive additions to the Federal payroll." As a result, he "was convinced that the major portion of our added funds must be spent with industry, education and other non-profit institutions." For Webb, the incredible magnitude of Project Apollo necessitated giving work to outside sources in order to accomplish the lunar landing within Kennedy's mandated timetable.[68]

By the mid-1960s, with the space program in full swing, NASA was directing over $4 billion per year to supporting organizations. More than 375,000 contract employees worked on NASA programs. Though widely distributed, the largest portion of funds went to contractors in sunbelt states that were growing economically. California, Louisiana, Alabama, Florida, and Texas received substantial shares.[69] These states provided strong support for NASA as the space program got underway. In the 1963 effort to cut $519 million from the NASA budget, not a single senator from those states supported the move. New York State also ranked high in the distribution of NASA contracts. Not surprisingly, neither of the New York senators voted to cut NASA's budget.

In his chapter on the Johnson years, Robert Dallek argues that Johnson promoted space expenditures as a means to stimulate the economies of the sunbelt states, particularly in the deep South. Johnson also had a second purpose in mind. He needed the votes of conservative southern Democrats for his Great Society goals. Space contracts helped to soften conservative opposition to government largesse. Tied as it was to cold war politics and economic growth, space spending was an easier vote for southern conservatives than a vote for welfare programs or civil rights. As Johnson himself remarked, it was easier for southern conservatives to accept Great Society initiatives once they had voted funds for an aggressive space program. Space spending helped to create a precedent for big government spending among fiscal conservatives. In that sense, space spending fit into Johnson's vision of the social transformation necessary to achieve his Great Society goals.[70]

Government contracts played a central role in Nixon's decision to approve the space shuttle program. The U.S. aerospace industry had been hard hit by the wind-down of Project Apollo and the cancellation of U.S. participation in the effort to develop a supersonic transport. Another major defeat in the government procurement of aerospace products would have hurt Republicans running in the 1972 elections. Nixon was concerned about "battleground states" like California and wanted his party to break into the solidly Democratic South. Ehrlichman, one of Nixon's advisors, recalled that the list of key states was quite short but "when you look at employment numbers [for the aerospace industry], and you key them to the battleground states, the space program has an importance out of proportion to its budget." While the space shuttle would not generate many jobs relative to overall employment, it would help in regions where Nixon's political fortunes lay. "You must not underemphasize that employment element," Ehrlichman said, "in Nixon's decision on the whole manned space program."[71]

As the space station issue moved to center stage, so did the implications of economic benefits. By 1992, the space station had created an estimated 75,000 jobs in more than half of the states, a point that NASA officials emphasized as they distributed maps revealing the distribution of spending during the congressional debate on NASA funding.[72] Representative David Obey of Wisconsin, an opponent of the program, complained that "there is no bigger pork item in the domestic budget than this item." Speaking in support of continued funding, Texas Representative Tom DeLay argued that it was virtually impossible to "deprive your state and your constituents of this important source of jobs and revenue." Only two members of the 27–person Texas congressional delegation voted to kill the program. The space station received similarly strong support from Alabama and southern California, both beneficiaries of significant space program largesse. When the issue reached the Senate, Barbara Mikulski of Maryland gave it strong support. Chair of the appropriations subcommittee handling the NASA budget, she voted with liberals on most issues but not on the space station. The space station program, she argued, would save jobs in the aerospace industry. "We are going to generate jobs today and jobs tomorrow," she argued, rebuffing efforts to cut the program from the federal budget. Mikulski expressed pride in the work of NASA's Goddard Space Flight Center, located in her home state, and noted that Maryland industries had received close to $18 million in space station contracts.[73]

As dramatic as the influence of contracting appears, it is not the primary force motivating congressional responses to space initiatives. Ideology is the strongest predictor of congressional voting patterns. Especially in recent years, party affiliation ranks a close second. The influence of economic benefit is diluted considerably by the tendency of conservative lawmakers to vote for the space program regardless of whether or not their constituents receive a great deal of funds from NASA. The tendency to switch is most pronounced among legislative liberals, for whom economic benefit creates a tendency to support programs with which they disagree ideologically.[74]

Conclusion

Presidents exercise leadership in the U.S. space program within a complex web of ideological, partisan, and economic constraints. Other nuances too subtle to discuss here, such as a personal infatuation with space exploration, have also moved individual political leaders to support or oppose presidential initiatives in space. Presidential leadership is not easy, given these constraints, nor is success assured. In this regard, the space program resembles other political issues. It is not as unique as its adherents would like to believe.

Although opportunities for presidential control are limited, the president continues to be the person who initiates the national space agenda. Before an initiative has any chance of political success, it must be endorsed by the president. The initiative may be modified substantially by the Congress, but it will not be considered to any great extent unless the president proposes it. No one but the president could have effectively initiated the race to the Moon in 1961, the space shuttle program in 1972, or the space station and space exploration initiatives of the 1980s. Leadership had to emanate from the White House, even if other politicians disagreed. As such, the president has been and continues to be the crucial player in the effort to define the overall space program. Without the president, no large-scale project could be placed on the national political agenda.

Once proposed, however, any large space effort has to be nurtured through the political process. For projects involving space technology, this takes place over many years. That process in Washington has never been particularly tidy, nor can it ever be in a pluralistic system. There may be fundamental agreement on overarching goals, such as the desire to participate in space exploration, but specific means for achieving those goals are constantly reevaluated and altered.

This constant reevaluation gives rise to what some observers have characterized as "heterogeneous engineering," a situation in which technology and politics develop simultaneously.[75] The image of presidential leadership created by Kennedy and Project Apollo is certainly attractive, but it does not abolish the general rule. Space policy is constructed within a complex web of institutions and interests that makes space exploration no more or less exceptional than other governmental activities.

Notes

1. See Arthur M. Schlesinger Jr., *The Imperial Presidency* (Boston: Houghton Mifflin, 1973).

2. Kennedy, "Presidential Address on Urgent National Needs," May 25, 1961, *Public Papers of the Presidents of the United States: John F. Kennedy, 1961* (Washington, D.C.: Government Printing Office, 1962), pp. 401–5.

3. John M. Logsdon, "An Apollo Perspective," *Astronautics and Aeronautics,* Dec. 1979, pp. 112–17, quotation is from p. 115.

4. Jane Van Nimmen and Leonard C. Bruno, with Robert L. Rosholt, *NASA Historical Data Book,* vol. 1, *NASA Resources, 1958–1968* (Washington, D.C.: NASA [SP-4012], 1988), pp. 137–41, 134, 63–119.

5. "Kennedy Box Score," *Congressional Quarterly Almanac, 1961* (Washington, D.C.: Congressional Quarterly, 1962), pp. 91–102.

6. Logsdon, *The Decision to Go to the Moon: Project Apollo and the National Interest* (Cambridge, Mass.: MIT Press, 1970), p. 129.

7. Tom Horton, "On Environment: If America Could Send a Man to the Moon, Why Can't We Clean Up Chesapeake Bay?" *Baltimore Sun,* July 22, 1984.

8. See, for example, George M. Low, Team Leader, to Richard Fairbanks, Director, Transition Resources and Development Group, "Report of the NASA Transition Team," Dec. 19, 1980, NASA Historical Reference Collection, NASA History Office, Washington, D.C.

9. Hans Mark, *The Space Station: A Personal Journey* (Durham, N.C.: Duke University Press, 1987), p. 36, see also p. 50.

10. This has been demonstrated too many times to be seriously questioned. See Walter A. McDougall, . . . *The Heavens and the Earth: A Political History of the Space Age* (New York: Basic Books, 1985), pp. 141–235; Logsdon, *Decision to Go to the Moon;* Harvey Brooks, "Motivations for the Space Program: Past and Future," in *The First 25 Years in Space: A Symposium,* ed. Allan A. Needell (Washington, D.C.: Smithsonian Institution Press, 1983), pp. 3–26; Rip Bulkeley, *The*

Sputniks Crisis and Early United States Space Policy: A Critique of the Historiography of Space (Bloomington: Indiana University Press, 1991).

11. Robert A. Divine, "Lyndon B. Johnson and the Politics of Space," in *The Johnson Years*, vol. 2, *Vietnam, the Environment, and Science*, ed. Robert A. Divine (Lawrence: University Press of Kansas, 1987), pp. 217–53.

12. "Success Rate History," *Congressional Quarterly Weekly Report* (CQWR), Dec. 19, 1992, p. 3896; see also William J. Keefe, *Congress and the American People* (Englewood Cliffs, N.J.: Prentice-Hall, 1988), chap. 4.

13. This took place on Aug. 20, 1958. See Loyd S. Swenson Jr., James M. Grimwood, and Charles C. Alexander, *This New Ocean: A History of Project Mercury* (Washington, D.C.: NASA [SP-4201], 1966), pp. 101–6.

14. "Congress Authorizes $3.4 Billion for Space Agency," *Congressional Quarterly Almanac, 1970* (Washington, D.C.: Congressional Quarterly, 1971), p. 204.

15. On these programs, see Bevan M. French and Stephen P. Maran, eds., *A Meeting with the Universe: Science Discoveries from the Space Program* (Washington, D.C.: NASA [EP-177], 1981); Paul A. Hanle and Von Del Chamberlain, eds., *Space Science Comes of Age: Perspectives in the History of the Space Sciences* (Washington, D.C.: Smithsonian Institution Press, 1981); Joseph N. Tatarewicz, *Space Technology and Planetary Astronomy* (Bloomington: Indiana University Press, 1990); Robert W. Smith, *The Space Telescope: A Study of NASA, Science, Technology, and Politics* (New York: Cambridge University Press, 1989); Karl Hufbauer, *Exploring the Sun: Solar Science since Galileo* (Baltimore: Johns Hopkins University Press, 1991).

16. "Major Legislation—Appropriations," *Congressional Quarterly Almanac, 1963* (Washington, D.C.: Congressional Quarterly, 1964), p. 170.

17. Ibid., p. 170.

18. Adam L. Gruen, "Deep Space Nein? The Troubled History of Space Station *Freedom*," *Ad Astra*, May/June 1993, pp. 18–23; Gruen, "The Port Unknown: A History of the Space Station Freedom Program" (unpublished manuscript, 1993), pp. 433–41, NASA Historical Reference Collection.

19. The following table summarizes the votes of continuing members of the House who changed their votes on the space station appropriation bill between 1992 and 1993. A "no" vote indicates that the member switched his or her vote from yes to no; a "yes" vote indicates that the member switched from no to yes. In each case, a "no" was a vote in favor of the space station and the president's position. The result is statistically significant for a one-tailed test at the .05 level ($x^2 = 2.9$):

	No	Yes	Total
Democrats	13	10	23
Republicans	5	12	17
	18	22	40

20. On these decisions, see Logsdon, "The Space Shuttle Program: A Policy Failure," *Science*, May 30, 1986, pp. 1099–105; Logsdon, "The Decision to Develop the Space Shuttle," *Space Policy* 2 (May 1986): 103–19; Howard E. McCurdy, *The Space Station Decision: Incremental Politics and Technological Choice* (Baltimore: Johns Hopkins University Press, 1990).

21. Synthesis Group, (Thomas P. Stafford, chair), *America at the Threshold: Report of the Synthesis Group on America's Space Exploration Initiative* (Washington, D.C.: Government Printing Office, 1991); "Space Program Faces Costly, Clouded Future," CQWR, Apr. 5, 1986, p. 732; "NASA Cuts Slow Ambitious Plans," *Congressional Quarterly Almanac, 1990* (Washington, D.C.: Congressional Quarterly, 1991), p. 435; "Bush Goes on the Counterattack against Mars Mission Critics," CQWR, June 23, 1990, p. 1958.

22. Leon E. Panetta, "Who Will Pay for Bush's 'Vision'?" *Washington Post*, Aug. 3, 1989, p. A27.

23. "Ticket to Space," *Washington Post*, July 23, 1989, p. D6.

24. "Bush Goes on the Counterattack," p. 1958.

25. National Academy of Sciences, Space Science Board, *Space Research Directions for the Future*, Publication 1403 (Washington, D.C.: National Academy of Sciences and National Research Council, 1966). For a more complete treatment of Project Voyager, see Edward C. Ezell and Linda N. Ezell, *On Mars: Exploration of the Red Planet, 1958–1978* (Washington, D.C.: NASA [SP-4212], 1984).

26. John E. Naugle, "Goals in Space Science and Applications," *Nuclear News*, Jan. 1969; Naugle, *First among Equals: The Selection of NASA Space Science Experiments* (Washington, D.C.: NASA [SP-4215], 1991).

27. W. David Compton and Charles D. Benson, *Living and Working in Space: A History of Skylab* (Washington, D.C.: NASA [SP-4208], 1983), pp. 101–2; *Aeronautics and Space Report of the President, 1989–1990 Activities* (Washington, D.C.: Government Printing Office, 1991), p. 161.

28. David Rubashkin, "Who Killed Grand Tour?" (unpublished manuscript, Apr. 30, 1993, American University Department of History), NASA Historical Reference Collection.

29. Mark Washburn, *Distant Encounters* (New York: Harcourt, Brace, Jovanovich, 1983); Linda Neuman Ezell, *NASA Historical Data Book*, vol. 3, *Programs and Projects, 1969–1978* (Washington, D.C.: NASA [SP-4012], 1988), pp. 221–28.

30. National Advisory Committee for Aeronautics to Killian's Office, White House, Aug. 6, 1958, NASA Historical Reference Collection.

31. G. Mennen Williams, quoted in William E. Burrows, *Deep Black: Space Espionage and National Security* (New York: Random House, 1987), pp. 94–95. See also Derek W. Elliott, "Finding an Appropriate Commitment: Space Policy under Eisenhower and Kennedy" (Ph.D. diss., George Washington University, 1992).

32. Glennan to Eisenhower, Dec. 28, 1960; Eisenhower to Glennan, Dec. 29,

1960, both in NASA Historical Reference Collection; "NASA Post Is Resigned by Glennan," *Baltimore Sun,* Dec. 30, 1960.

33. Eisenhower, "Are We Headed in the Wrong Direction?" *Saturday Evening Post,* Aug. 11, 1962, p. 24.

34. Eisenhower, "Why I Am a Republican," *Saturday Evening Post,* Apr. 11, 1964, p. 19.

35. Richard E. Horner, Northrop Corp., to Glennan, June 1, 1961; Glennan to J. B. Lawrence, chair of International Fact Finding Institution, May 16, 1961; Glennan to Nixon, June 14, 1961; Glennan to Eisenhower, June 14, 1961, and Nov. 13, 1961; Glennan to George Kistiakowsky, Dec. 4, 1961; Glennan to Killian, Dec. 4, 1961; Killian to Glennan, Dec. 19, 1961; Glennan to Neil McElroy, Sept. 22, 1961, all in Glennan Personal Papers, 19DD4, Archives, Case Western Reserve University, Cleveland, Ohio.

36. Glennan to Eisenhower, May 31, 1961, Glennan Personal Papers, 19DD4.

37. "Cut in NASA Funds Highlights Independent Offices," *Congressional Quarterly Almanac, 1963,* p. 170.

38. "Senate Rejects Space Fund Cuts," *Congressional Quarterly Almanac, 1967* (Washington, D.C.: Congressional Quarterly, 1968), p. 29-S.

39. "Breakdown of Party Unity Votes," CQWR, Dec. 28, 1991, p. 3789.

40. James M. Lindsay, "Congress and Defense Policy, 1961–1986," *Armed Forces and Society* 13 (Spring 1987): 371–401; Keefe, *Congress and the American People,* pp. 122–28.

41. "Breakdown of Party Unity Votes," CQWR, Dec. 18, 1993, p. 3480.

42. On this issue, see Aaron Wildavsky, *The Rise of Radical Egalitarianism* (Washington, D.C.: American University Press, 1991).

43. On these issues, see Doris Kearns, *Lyndon Johnson and the American Dream* (New York: Harper and Row, 1976); Christopher Lasch, *The Culture of Narcissism: American Life in an Age of Diminishing Expectations* (New York: W. W. Norton, 1979); Bruce Miroff, *Pragmatic Illusions: The Presidential Politics of John F. Kennedy* (New York: David McKay, 1976).

44. See R. Cargill Hall, "The Origins of U.S. Space Policy: Eisenhower, Open Skies, and Freedom of Space," in *Exploring the Unknown: Selected Documents in the History of the U.S. Civil Space Program,* vol. 1: *Organizing for Exploration,* ed. John M. Logsdon, with Linda J. Lear, Jannelle Warren-Findley, Ray A. Williamson, and Dwayne A. Day (Washington, D.C.: NASA [SP-4218], 1995), pp. 213–29; Robert A. Divine, *The Sputnik Challenge: Eisenhower's Response to the Soviet Satellite* (New York: Oxford University Press, 1993); Constance McLaughlin Green and Milton Lomask, *Vanguard: A History* (Washington, D.C.: Smithsonian Institution Press, 1971).

45. James R. Killian Jr., *Sputnik, Scientists, and Eisenhower: A Memoir of the First Special Assistant to the President for Science and Technology* (Cambridge, Mass.: MIT Press, 1977), p. 68.

46. Ibid.; John Prados, *The Soviet Estimate: U.S. Intelligence Analysis and*

Russian Military Strength (New York: Dial Press, 1982), p. 60; Stephen E. Ambrose, *Eisenhower*, vol. 2, *The President* (New York: Simon and Schuster, 1984), p. 257.

47. Swenson et al., *This New Ocean*, pp. 101–12.

48. See Bulkeley, *Sputniks Crisis*.

49. Glennan to Webb, July 21, 1961, Glennan Personal Papers, 19DD4.

50. "Cut in NASA Funds," p. 172.

51. The ideological shift bears up statistically. See John Low, "Economic Benefit, Ideology, and NASA Voting in the U.S. Senate" (unpublished manuscript prepared for the American University, 1991, supplemented by an analysis of House voting patterns).

52. In January 1970, Thomas O. Paine, Nixon's appointee as the NASA administrator, described a somber meeting in which Nixon told him that both public opinion polls and political advisors indicated that the mood of the country supported drastic cuts in the space and defense programs. Paine, memorandum, "Meeting with the President, January 22, 1970," Jan. 22, 1970; Caspar W. Weinberger, interview by Logsdon, Aug. 23, 1977, both in NASA Historical Reference Collection.

53. Weinberger, interview.

54. John Ehrlichman, interview by Logsdon, May 6, 1983, NASA Historical Reference Collection. See also Low to Fletcher, "Items of Interest," Aug. 12, 1971; Fletcher to Jonathan Rose, Special Assistant to the President, Nov. 22, 1971, both in NASA Historical Reference Collection.

55. Ehrlichman, interview.

56. Jacob E. Smart, NASA assistant administrator for DOD and Interagency Affairs, to Fletcher, "Security Implications in National Space Program," with attachments, Dec. 1, 1971, James C. Fletcher Papers, Special Collections, Marriott Library, University of Utah, Salt Lake City; Fletcher to Low, "Conversation with Al Haig," Dec. 2, 1971, NASA Historical Reference Collection.

57. Quoted in Ken Hechler, *Toward the Endless Frontier: History of the [House] Committee on Science and Technology, 1959–79* (Washington, D.C.: House of Representatives, 1980), p. 274.

58. George D. Brown, "Priorities: Responding to a Changing Context," remarks at National Academy of Sciences, Jan. 24, 1992, p. 6, NASA Historical Reference Collection.

59. Kennedy, "If the Soviets Control Space . . . They Can Control Earth," *Missiles and Rockets*, Oct. 10, 1960, pp. 12–13; Kennedy, "Address at Rice University in Houston on the Nation's Space Effort," Sept. 12, 1962, *Public Papers of the Presidents of the United States: John F. Kennedy, 1962* (Washington, D.C.: Government Printing Office, 1963), pp. 668–71.

60. Kennedy, "Address at Rice University," p. 669.

61. White House, Office of the Press Secretary, "The White House, Statement by the President," Jan. 5, 1972, Richard M. Nixon Presidential Files, NASA Historical Reference Collection.

62. Reagan, "Address before a Joint Session of Congress on the State of the Union," Jan. 25, 1984, *Public Papers of the Presidents of the United States: Ronald Reagan, 1984* (Washington, D.C.: Government Printing Office, 1986), p. 90.

63. Frederick Jackson Turner developed the "Frontier Thesis" that guided inquiry into much of American history for a generation. It continues to inform many popular images of the American West by traditional America. Turner outlined the major features of the thesis in *The Frontier in American History* (New York: Holt, Rinehart, and Winston, 1920), which included the seminal 1893 essay, "The Significance of the Frontier in American History."

64. Wildavsky, *Rise of Radical Egalitarianism.*

65. Patricia Nelson Limerick, "The Final Frontier?" *Wilson Quarterly* 14 (Summer 1990): 82–83, quotation is from p. 83. See also Richard Slotkin, *Gunfighter Nation: The Myth of the Frontier in Twentieth-Century America* (New York: Atheneum, 1992); Ray A. Williamson, "Outer Space as Frontier: Lessons for Today," *Western Folklore* 46 (Oct. 1987): 255–67; M. Jane Young, "'Pity the Indians of Outer Space': Native American Views of the Space Program," *Western Folklore* 46 (Oct. 1987): 269–79; Claire R. Farrer, "On Parables, Questions, and Predictions," *Western Folklore* 46 (Oct. 1987): 281–93; Stephen J. Pyne, "Space: A Third Great Age of Discovery," *Space Policy* 4 (Aug. 1988): 187–99. On the limitations of analogy in historical study, see Bruce Mazlish, ed., *The Railroad and the Space Program: An Exploration in Historical Analogy* (Cambridge, Mass.: Harvard University Press, 1965); Richard E. Neustadt and Ernest R. May, *Thinking in Time: The Uses of History for Decision Makers* (New York: Free Press, 1986).

66. Linda R. Cohn and Roger G. Noll, with Jeffrey S. Banks, Susan A. Edelman, and William M. Pegram, *The Technology Pork Barrel* (Washington, D.C.: Brookings Institution Press, 1991).

67. "He Did It All," *CWRU Magazine*, Feb. 1990, p. 14. On the creation of the Houston center, see Henry C. Dethloff, *"Suddenly Tomorrow Came . . .": The First Steps in Human Space Flight, Johnson Space Center, 1957–1990* (Washington, D.C.: NASA [SP-4307], 1993).

68. Glennan, *The Birth of NASA: The Diary of T. Keith Glennan*, ed. J. D. Hunley (Washington, D.C.: NASA [SP-4105], 1993), pp. 5, 120; Arnold L. Levine, *Managing NASA in the Apollo Era* (Washington, D.C.: NASA [SP-4102], 1982), pp. 65–72.

69. NASA, *Annual Procurement Report Fiscal Year 1965* (Washington, D.C.: NASA, 1965), p. 40.

70. McDougall, . . . *The Heavens and the Earth*, pp. 322–23, 376, 389–96; Kennedy to Johnson, July 29, 1963; Johnson to Kennedy, July 31, 1963; Newton Minow, interview, all in White House Famous Names, Lyndon Baines Johnson Library, Austin, Texas.

71. Ehrlichman, interview. This aspect of the issue was also brought home to Nixon by other factors such as letters and personal meetings. See Frank Kizis to Nixon, Mar. 12, 1971; Noble M. Melencamp, White House, to Kizis, Apr. 19,

1971, both in RG 51, series 69.1, box 51–78–31, National Archives and Records Administration.

72. NASA, *Space Station Freedom Media Handbook* (Washington, D.C.: NASA, 1992), p. 87.

73. "VA, HUD, Agencies' Plates Filled with Pork," *Congressional Quarterly Almanac, 1992* (Washington, D.C.: Congressional Quarterly, 1993), p. 643; "*Freedom* Thwarts Funding Foes; But Bigger Budget War Ahead," CQWR, May 2, 1992, p. 1157; "*Freedom* Fighters Win Again: Senate Keeps Space Station," CQWR, Sept. 12, 1992, p. 2722.

74. Low, "Economic Benefit, Ideology, and NASA Voting."

75. John Law, "Technology and Heterogeneous Engineering: The Case of Portuguese Expansion," pp. 111–34; and Donald MacKenzie, "Missile Accuracy: A Case Study in the Social Processes of Technological Change," pp. 195–222, both in *The Social Construction of Technological Systems: New Directions in the Sociology and History of Technology,* ed. Wiebe E. Bijker, Thomas P. Hughes, and Trevor J. Pinch (Cambridge, Mass.: MIT Press, 1987).

Contributors

MICHAEL R. BESCHLOSS is well known as a political commentator in Washington, D.C., and has written extensively on the Kennedy administration and foreign policy, including *Mayday* (Harper and Row, 1986); *The Crisis Years: Kennedy and Khrushchev, 1960–1963* (Harper, 1991); and, with Strobe Talbott, the widely acclaimed *At the Highest Levels: The Inside Story of the End of the Cold War* (Little, Brown, 1993). Beschloss is also a commentator for National Public Radio, CNN, and other media, specializing in analysis of the history of the recent American presidency and American foreign relations.

DAVID CALLAHAN is a graduate student working with Fred I. Greenstein at Princeton University. He is the author of *Dangerous Capabilities: Paul Nitze and the Cold War* (HarperCollins, 1990).

ROBERT DALLEK is a professor of history at the University of California, Los Angeles. He has published widely on American foreign policy and the modern presidency and is the author of several books on recent American history, including *Franklin D. Roosevelt and American Foreign Policy, 1932–1945* (Oxford University Press, 1979), which received the Bancroft Prize from the Organization of American Historians. He is working on a two-volume biography of Lyndon Johnson, the first volume of which, *Lone Star Rising: Lyndon Johnson and His Times, 1908–1960* (Oxford University Press, 1991), has already appeared.

FRED I. GREENSTEIN is a professor of politics and chair of the Program in Leadership Studies at the Woodrow Wilson School of Public and International Affairs, Princeton University. The principal architect of the revisionist treatment of Dwight D. Eisenhower as a political leader, which began in the 1980s with the publication of his pathbreaking study *The Hidden-Hand Presidency: Eisenhower as Leader* (Basic Books, 1982), he is also the editor of *Leadership in the Modern Presidency* (Harvard University Press, 1988).

ROBERT H. FERRELL has taught American history at Indiana University, Bloomington, for more than thirty years. One of the deans of the history of U.S. foreign relations, Ferrell has specialized in twentieth-century U.S. diplomatic relations and the role of the American presidency in shaping world affairs. He has written or edited twenty-nine books, including *American Diplomacy: The Twentieth Century* (W. W. Norton, 1987); *America in a Divided World, 1945–1972* (University of South Carolina Press, 1975); *American Diplomacy* (W. W. Norton, 1975); and *Ill-Advised: Presidential Health and Public Trust* (University of Missouri Press, 1992). His most recent book is the widely acclaimed *Harry S. Truman: A Life* (University of Missouri Press, 1994).

JOAN HOFF is a professor of history at Indiana University, Bloomington, and during 1992–93 held the Mary Ball Washington Chair in American History at University College, Dublin. One of the most respected individuals working in modern American and women's history, she has published, among other books and essays, *American Business and Foreign Policy, 1920–1933* (University Press of Kentucky, 1971); "Richard M. Nixon: The Corporate Presidency," in *Leadership in the Modern Presidency*, ed. Fred I. Greenstein (Harvard University Press, 1988); and a full-length biography entitled *Nixon Reconsidered* (Basic Books, 1994).

ROGER D. LAUNIUS is chief historian of the National Aeronautics and Space Administration, Washington, D.C. He has published articles on the history of aeronautics and space in several journals and has written or edited twelve books, including *Joseph Smith III: Pragmatic Prophet* (University of Illinois Press, 1988); *Missouri Folk Heroes of the Nineteenth Century* (Independence Press, 1989); with Linda Thatcher, *Differing Visions: Dissenters in Mormon History* (University of Illinois Press, 1994); and *NASA: A History of the U.S. Civil Space Program* (Robert E. Krieger, 1994).

JOHN M. LOGSDON is director of the Space Policy Institute at George Washington University and a recognized authority on the evolution of space policy since *Sputnik*. He is the author of the seminal study *The Decision to Go to the Moon: Project Apollo and the National Interest* (MIT Press, 1970) and a large number of space policy articles appearing in many publications, especially *Space Policy*, of which he is the North American editor.

HOWARD E. McCURDY is a professor of public affairs at American University, Washington, D.C. He is the author of numerous articles on public policy and governmental management and books on the U.S. space program, including *The Space Station Decision: Incremental Politics and Technological Choice* (Johns Hopkins University Press, 1990) and *Inside NASA: High Technology and Organizational Change in the U.S. Space Program* (Johns Hopkins University Press, 1993).

LYN RAGSDALE is a professor of political science at the University of Arizona, Tucson. One of the most respected of the new generation of scholars working in the field of the modern presidency, she is the author of *Presidential Politics* (Houghton Mifflin, 1993) and, with Gary King, *The Elusive Executive: Discovering Statistical Patterns in the Presidency* (CQ Press, 1988).

Index